"As a physicist I was pleased to see the ideas of physics given such a positive exposure by a popular author. Martina Sprague presents clear and simple explanations of the basic principles of physics, and she shows in a logical and persuasive way how understanding these principles can enhance a fighter's performance. She nicely explains each idea and presents concrete examples of its application. But, even more than this, she gives one such a nice sense of the principles, that correct form, in terms of power and speed, is generated naturally and intuitively.

I do not practice martial arts. However I do play tennis, and I am always eager to improve my technique. I had no intention of getting any advantage in my sport by reading this book, and yet I did just that. The next day, after reading chapter four (on "Direction") I found that it applied to my tennis service motion. I could sense that my body weight was not in line with my service arm swing. And this naturally reduced the power of my serve. This was the same principle which the text applied to the power of a punch or kick. I was quite surprised by the advantage I had gained from reading the book. Certainly other sports, such as baseball and golf, work on the same principles."

— *Steve Landy*
adjunct assistant professor of physics, IUPUI

Fighting Science
The Laws of Physics for Martial Artists

Fighting Science
The Laws of Physics for Martial Artists

by Martina Sprague

 Turtle Press   Hartford

FIGHTING SCIENCE: THE LAWS OF PHYSICS FOR MARTIAL ARTISTS Copyright © 2002 Martina Sprague. All rights reserved. Printed in the United States of America. No part of this book may be reproduced without written permission except in the case of brief quotations embodied in articles or reviews. For information, address Turtle Press, PO Box 290206, Wethersfield CT 06129-0206.

To contact the author or to order additional copies of this book:
    Turtle Press
    P.O. Box 290206
    Wethersfield, CT 06129-0206
    1-800-77-TURTL

ISBN 1-880336-72-3
LCCN 2002005253

Printed in the United States of America

10 9 8 7 6 5 4 3 2

Library of Congress Cataloging-in-Publication Data

Sprague, Martina.
  Fighting science : the laws of physics for martial artists / by Martina Sprague.
    p. cm.
Includes index.
  ISBN 1-880336-72-3
  1. Martial arts. 2. Physics. I. Title.
  GV1101 .S68 2002
  796.8—dc21
                              2002005253

Dedication

To my favorite instructor, Keith Livingston, for the many years of pain and glory.

Acknowledgments

A special thank-you to Tom, my soul mate and husband of eighteen years, who is as interested in physics as I am, and who let me bounce my ideas off of him. Also, thank-you to my martial arts students, who have provided me with so many opportunities to think about and analyze techniques and concepts. And, finally, thank-you to my first physics instructor in Sweden, Åke Johansson, who was instrumental in helping me develop a love for physics, and who sparked my interest from the beginning by asking why the forest, when seen through a glass of water, is upside down.

Table Of Contents

Preface	**15**
Introduction	**19**
Physics And Strategy: An Introduction To The Fighting Concepts	**21**
Ten Fundamental Fighting Concepts	22
Use Stronger Tools Against Weaker Tools	23
Use Techniques That Can Be Escalated	25
Eliminate Several Of Your Opponent's Weapons Simultaneously	27
Use Your Sense Of Touch	29
Use Several Weapons Together	30
Use Shock Value To Split Your Opponent's Focus	31
Rely On Superior Defense	32
Rely On Speed And Surprise	34
Use The Inferior Position As A Strength	35
Use Logical Sequencing Of Technique	36
Indomitable Spirit	37
Center Of Gravity: Balance, How Important Is It?	**39**
Balance And Stance	40
Center Of Gravity	40
The Power Of The Lead Leg Kick	47
Balance In The Hard And Soft Styles	49
Balance Manipulation In Throws	50
Balance Manipulation In Joint Locks And Takedowns	52
Summary And Review	59
Balance In Stances	60
Balance When Striking	62
Balance When Kicking	63
Balance In Defense	66
Balance When Moving	71
Balance In Throws And Takedowns	73
Balance When Grappling	75
Balance Quiz	77
Glossary	79
Momentum: Without Movement, Nothing Happens	**81**
Body Mass In Motion	82
Working With Momentum	85
A Different Look At Distance And Reach	90
Speeding Up Your Punches	95
The Power Of The Sideways Stance	98
Adding Momentums	99

Summary And Review	101
Momentum When Striking	101
Momentum When Kicking	106
Momentum In Defense	112
Momentum In Throws And Takedowns	116
Momentum When Grappling	120
Mass And Momentum Quiz	122
Glossary	124

Direction: F=ma, An Unbeatable Combination — **127**

Speed In Combinations	128
The Resultant Force	133
The Dangers Of Exhausting The Motion	137
The Overhand Projectile	138
Summary And Review	140
Direction When Striking	140
Direction When Kicking	143
Direction In Defense	146
Direction In Throws And Takedowns	147
Direction When Grappling	149
Inertia And Vector Quiz	151
Glossary	153

Rotational Speed And Friction: Circular Movement = Power — **155**

Jump Kicks	156
Jumping, Sliding, And Stepping	157
Circular Motion	160
Rotation Of The Fist	162
Torque	163
The Catch-22	166
The Speed And Power Of Spinning Techniques	167
Rotational Inertia	171
Summary And Review	175
Speed When Striking	175
Speed When Kicking	179
Speed In Defense	180
Speed In Throws And Takedowns	180
Speed When Grappling	184
Friction And Rotational Speed Quiz	187
Glossary	189

Impulse: Striking Through The Target — **191**

Decreasing The Time Of Impact In The Grappling Arts	192
Decreasing The Time Of Impact In The Striking Arts	196
Reversing Direction By "Bouncing"	198

Impulse When Kicking	201
Three Types Of Impact	202
Impulse Exercise	204
Summary And Review	207
Impulse When Striking	207
Impulse When Kicking	209
Impulse In Defense	211
Impulse In Throws And Takedowns	212
Impulse When Grappling	213
Impulse Quiz	215
Glossary	217
Conservation Of Energy: The Workload	**219**
What Is Power?	220
The Physics Of The One-Inch Punch	221
Conservation Of Energy	222
Using The Spinning Back Kick Strategically	225
Summary And Review	228
Conserving Energy When Striking	228
Conserving Energy When Kicking	230
Conserving Energy In Defense	232
Conserving Energy In Throws And Takedowns	234
Conserving Energy When Grappling	235
Energy Quiz	237
Glossary	239
***Ki*-netic Energy: Mind Over Matter**	**241**
The Power Paradox	242
How Important Is Kinetic Energy?	244
Mind Over Matter	248
Flash Knockout	250
Other Incredible Physical Feats	251
The Strength Of The Arch	253
Imposing Your Will	254
Summary And Review	255
Kinetic Energy Quiz	258
Glossary	259
Conclusion	**261**

Preface

You hate physics? You're just not a math whiz? Many people squirm when they hear the word *physics*, and the first thing that comes to mind are numbers and letters mixed into some sort of incomprehensible language called *equations*. Well, don't worry! The physics we will discuss is conceptual physics, which relies mainly on concepts rather than equations. Concepts are ideas that the reader is already familiar with. These ideas are then related to martial arts (power, in particular). The following equations, which you will see in the text, are there only to strengthen the concepts.

1. Momentum = mass (weight) X velocity (speed)
2. Force = mass X acceleration
3. Torque = lever arm X force
4. Impulse = force X time
5. Work = force X distance
6. Power = work/time
7. Kinetic energy = 1/2 mass X velocity squared

The equations will be embedded in the text and, where appropriate, will appear inside of a box above a picture relating to the equation. The equations will also use different *size* lettering to show which component part of the equation is strongest, as shown on the following page.

16 Fighting Science

**Change in momentum = force X time
or
Impulse = Ft**

In the equation **Impulse = Ft**, a small size **F** indicates a small force, and a large size **t** indicates a long time (left picture), whereas a large size **F** indicates a large force, and a small size **t** indicates a short time (right picture). This has been done to help the reader visualize what is happening through the pictures and text, through the equation in the box, and through the different size lettering in the equation above the picture. Note that different *size* lettering is not the same as upper or lower case, and is used only to emphasize which part of the equation is most significant to power. **F** will always be upper case, and **t** will always be lower case, simply because that's how it is written for consistency in the world of physics.

Also note that certain words that have an exact meaning in physics have occasionally been used in a more everyday type of language. An example is *power*. In physics, this is defined as work/time. But to the martial artist, power takes on a different meaning, and is commonly used to determine how much damage one is able to do when landing a strike. The way the martial artist uses the word power might be disturbing to the student of physics. But it should be borne in mind that the book is written primarily for the student of martial arts. For the purpose of this book, power should be thought of as the force of impact of a punch or kick. Power is also described as a vector, but for this to hold true in physics, it would be more appropriate to substitute power with force. A student of physics might also frown on the fact that I have used only numbers in most of the equations, without specifying the units. This has been done for simplification purposes, as the book focuses on concepts rather than equations. To tell a martial arts student that he should strike with a force of a certain number of *newtons*, would have meaning only if he had some prior knowledge of physics.

Once you truly understand the principles of balance, body mass in motion, inertia, direction, rotational speed, friction, impulse, and kinetic energy, the need to memorize hundreds of techniques vanishes. A true principle applies to all techniques and all people, whether you are standing, kneeling, prone, or supine, whether you are big or small, strong or weak. Physics, in itself, is neither good nor bad. It can neither be given to you, nor taken away; it applies equally to all people at all times. It's how you use it that makes the difference. I am confident that you will find the concepts of physics and martial arts discussed in this book enjoyable and easy to understand and apply.

Introduction

Martial arts means the *intricate study of combat*, and the purpose of this book is to analyze, according to the principles of physics, one of the most important assets to successful fighting: *power*. It has been said that a successful martial artist doesn't need size or strength, because "it is all in the *technique*." It has also been said that the power of a martial artist seems to increase quickly with weight, and that the best lightweight fighter in the world will be defeated every time by an unranked heavyweight in a bar brawl.

So, is it size and physical strength that makes you the winner, or is it experience and dedication to correct technique?

I once knew a martial artist who claimed that his instructor had such terrific powers that he could strike you from across the room without being within reach to physically touch you. I had at that time studied martial arts on a daily basis for about seven years and, being the down to earth type of person I was, this martial artist's claim did not only seem ridiculous; it did not even raise an ounce of curiosity in me. We have all heard stories of such feats a hundred times, but we have yet to meet those who can support them. Because I respected this martial artist as a good person and excellent sparring partner, I let the issue go and did not comment on it further. But as the years went by, I was intrigued by what had made him and so many others in the martial arts support this "mind over matter" type of fighting. After giving the issue substantial thought, and after disregarding possible differences in semantics, I found that it is, in fact:

- **Possible** to strike a person from a distance without physically touching him, but only if he knows in advance that he is going to get struck.

- **Not possible** to strike a person from a distance without physically touching him, if he uses his own "powers" to counteract the blow.

And then there are martial artists who walk barefoot on burning coal. Although, to the onlookers, it takes fortitude to perform the feat, I am not sure what exactly it is supposed to prove as far as martial arts is concerned. Intellectually, however, I understand the principles of physics behind these claims and others.

Most books about power in the martial arts rely on physical conditioning and prompt the reader to do push-ups and sit-ups and plyometrics. But I will attempt to take you through the back door and show you the *principles of physics* behind power. My purpose is not to negate the importance of physical conditioning, but rather to complement it by broadening your understanding of the laws of nature, the importance of correct technique, and how a smaller person can make these laws work to his advantage. Read on, and I will reveal to you the "secrets."

Physics And Strategy
An Introduction To The Fighting Concepts

This book is meant to take you to the highest stage of learning through the laws of physics. The first chapter will compare stand-up fights (karate, kick-boxing) and ground fights (grappling), and explore how the concepts of physics and strategy apply to both. In subsequent chapters, all the terms and concepts will be defined in detail and broken down into their component parts. I will then show you how to train using the laws of physics to your advantage.

Before you can gain proficiency as a fighter, you must learn proper mechanics of technique. This is called the *mechanical* stage, or learning by rote. Simply put, it is memorization without understanding. The mechanical stage will do you little good in actual sparring, yet it is needed to provide a foundation for continued growth.

The second stage of learning is called *understanding*. When you have reached this level, you will know why you do a particular technique and when, and why you do the moves in a particular sequence. You can now answer questions about the technique, but without necessarily being proficient in its execution.

The third stage of learning is called *application*. This is where you can use what you have learned in an unrehearsed sparring match.

The fourth stage of learning is called *correlation*. You can now see how the concepts for one technique can be applied to another, or how the concepts for stand-up fighting can be applied to grappling, and vice versa. Providing that you have learned sound mechanics of technique first, your knowledge from one art will now carry over to another, allowing you to diversify your skill without spending years perfecting a particular art. You can now become your own instructor.

Ten fundamental fighting concepts

The following concepts are not necessarily listed in their order of importance, nor are they the only important concepts in a fight. I have included an example of a specific technique for each concept. We will then discuss each concept in more detail. For the purpose of this section, it is also assumed that both you and your opponent fight empty handed, and that you fight one person at a time only.

1. Use stronger tools against weaker tools. For example, in a stand-up fight, use your elbow to block your opponent's round house kick. In a ground fight, use your leg to break a wrist grab.

2. Use techniques that can be escalated. In a stand-up fight, finishing strikes should follow set-up strikes. In a ground fight, a joint lock can be escalated to a dislocation technique, which can be escalated to a breaking technique.

3. Eliminate several of your opponent's weapons simultaneously. In a stand-up fight, attack when your opponent is at a disadvantaged position; for example, when he is on one leg and is unable to move away. In a ground fight, apply a joint lock while simultaneously pinning your opponent's head.

Note: The head is perhaps the most important body part to isolate. If your opponent can't move his head, the movement in the rest of his body will be severely limited.

4. Use your sense of touch. In a stand-up fight, when shoulder to shoulder with your opponent, any small move will be telegraphed through feel. In a ground fight, train with a blindfold to enhance your sense of touch.

5. Use several weapons together. In a stand-up fight, throw combinations. In a ground fight, lock your opponent's legs with your legs and execute a figure-four choke.

6. Use shock value to split your opponent's focus. In a stand-up fight, attack the same target multiple times. In a ground fight, grab soft tissue areas where sensitive nerves are located.

7. Rely on superior defense. In a stand-up fight, combine defense with offense. In a ground fight, continuously look for weakness.

8. Rely on speed and surprise, and the "DON'T WAIT--CREATE" principle. Be in charge. In a stand-up fight, initiate the attack. In a ground fight, be explosive.

9. Use the inferior position as a strength. In a stand-up fight, use strategy to reverse positions when cornered. In a ground fight, take advantage of your opponent's higher center of gravity when he is straddling you.

10. Use logical sequencing of techniques. In general, logical techniques are those that have a smooth flow with no awkward movements. Logic also involves distance and position.

Use stronger tools against weaker tools

Strength is a combination of the anatomical composition of your weapon and correct strategy (distance, timing, movement, etc). The elbow, for example, is a strong weapon because it is small and hard, and can be used to inflict considerable pain against the bony areas of your opponent's insteps, shins, or jaw line. Because of the small surface area of the elbow, it allows you to use a greater force per square inch than if you use the whole surface area of your hand. Because of the close proximity of the elbow to the body, you can rely on your body mass to generate power. You do this by keeping your elbow in front of your body and letting the movement of your strike originate in your body. Strategically, because the elbow is closer to your body than the hand, you can protect your midsection with your elbow and arm, simultaneously using the elbow as a striking block. By blocking your opponent's kick with your elbow, offense and defense is accomplished simultaneously. In addition, your opponent will experience a great deal of pain, which will make him fearful of throwing that kick again. You have now eliminated one of his effective weapons.

Note: Remember that the concept we are working on is how to use a *stronger weapon* against a *weaker weapon*, and not necessarily how to use your elbow against your opponent's shin. What other types of strong against weak techniques can you think of?

24   Fighting Science

Martina Sprague uses her stronger knife edge side kick against her opponent's weaker knee.

If your goal is to reach the highest level of learning; that of correlation, you must be able to apply the same concepts both to stand-up arts and ground arts. As long as you have a basic understanding of ground work and the anatomical limitations of the joints, this is true even if you have never studied ju-jutsu in particular. In a ground fight, strength usually comes through leverage (torque) and the understanding that a grip will break at its weakest point. When your opponent grabs your wrist, you can break the grip by bringing your stronger leg over the top of your opponent's weaker arm. In what other ways can you use strong against weak?

In grappling, you can use strong against weak by reinforcing your grip. When Martina's opponent grabs her wrist, she reinforces the grip with her free hand and forearm and drops her opponent to his knees.

Physics and Strategy

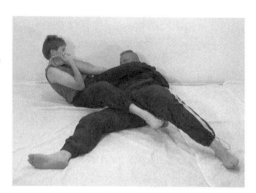

Here, Martina uses her stronger leg against her opponent's weaker neck, and her stronger arm against her opponent's weaker elbow.

Use techniques that can be escalated

Escalation, in fighting, means to increase the intensity. In a stand-up fight, the last technique should feel worse than all previous techniques. This can be done by throwing combinations that have a natural flow, and where the speed can build continuously throughout the combination. Natural flow means less start/stop movement and less need to overcome inertia. The greater speed of the last strike translates into power. Strategically, escalation of power is intended to end the fight. First, by striking the same target harder and harder, the anatomical limitation of that target will eventually give in to the force. Second, by striking multiple targets harder and harder, your opponent will feel overwhelmed and be unable to defend against the strikes. Striking multiple targets is likely to create openings for the knockout or finish.

In a stand-up fight, force can be escalated by going from a block, to a strike, to a joint lock.

26 Fighting Science

Again, the concept we are working is not how to build speed through combinations, but how to escalate the force, regardless of what type of techniques you use. In sports grappling, a technique should be escalated to the point of submission. First, gain a joint lock and apply just enough pressure to cause pain. If your opponent has a high pain threshold, escalate the technique into a possible dislocation of the joint (by now the referee should have intervened). The next stage is the actual breaking of the joint. A technique can be escalated by going against the natural movement of the joint. Because of torque and the anatomical limitations of the joint, doing damage now takes relatively little force, especially if the lever arm is long. Strategically, escalation of force is used to get pain compliance. At the earliest stage, there is no permanent damage to the joint, and as soon as the grip is released, the pain will subside. But knowing that a dislocation or break is imminent, is often enough to make a person submit.

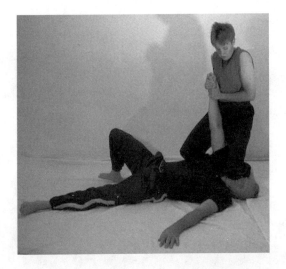

In a ground fight, force can be escalated by going from a knee press with control of the arm to a full arm bar.

Note: In a high threat street encounter, you may want to skip the intermediate dislocation stage and go into breaking right away.

Physics and Strategy

Eliminate several weapons simultaneously

Any time you use one of your weapons, it is momentarily tied up in that technique. When your opponent throws a punch, he is unable to simultaneously cover his ribs on the side of the punching arm. This gives you an automatic opening. In a ground fight, when your opponent grabs you with both hands, he is unable to simultaneously strike you. Providing that it is not your wrists he is grabbing, this leaves your hands free for striking or grabbing.

The fewer effective weapons your opponent has, the less diverse he will be, and the less you will have to worry about. In a stand-up fight, you can rely on the "completion of motion" principle, which states that *before another technique can be initiated, the motion of the first technique must first be completed.* This is especially beneficial whenever your opponent throws a kick, but can be used with punches as well. When your opponent kicks, the kicking leg will be tied up in the technique until the kick is complete. You will therefore know where that weapon is. In addition, because the legs are used for walking, your opponent won't be able to reposition himself until the motion of the kick is complete. This gives you a strategical advantage. In essence, you are tying up not only your opponent's kick (which he is using at the moment), but also his hands. Think about this: how beneficial is it to punch and kick simultaneously? Your strikes won't be very powerful, because your body must have the opportunity to reset before initiating a new technique.

Which weapons are being eliminated simultaneously in this sparring match? Answer: both the kick and the punches; the kick because of the leg check, and the punches because the fighter has his hands tied up around his opponent's neck.

When your opponent kicks, you can rely on his narrow foundation to unbalance him with a counter-attack. This makes it even more difficult for him to simultaneously use another weapon against you. Again, the principle is not about attacking when your opponent is on one leg, but about eliminating the use of several weapons. This can also be done through positioning. You can eliminate many of your opponent's techniques by working from an angle to the side. You can eliminate most of his techniques by positioning yourself behind him.

In a ground fight, several weapons can be eliminated simultaneously by pinning your opponent's head to the ground. The neck is an inherently weak area of our anatomy, and therefore relatively easy to control. When your opponent is on his back or stomach, turning his head to the side and placing your shin across his jaw line, will make it difficult for him to move or apply any type of effective defense. It can therefore be said that the person who controls the head, controls the fight. Applying a joint lock in a chaotic situation can be rather difficult, because it requires the use of fine motor skills. But once you have control of the head, you can take your time to properly apply a joint lock. You can even ask your opponent to position his hands for you: *"Place your left hand against the small of your back! Turn it palm up! Don't move!"* Increase the pressure on his head until he complies. Because both your hands are free and you have compliance, you can now use fine motor skills to apply the joint lock correctly.

When controlling the head, you are eliminating several of your opponent's weapons through the pounds per square inch principle (your shin against your opponent's jaw line). What is not as obvious is that you are also using leverage against your opponent's head, because his head is unable to turn a full 360 degrees. Once the head gets to its maximum turn, very little force is required to produce a high torque. Impulse is also increased because the neck has very little give. Your opponent is therefore controlled rather easily. Strategically, you are relying on pain compliance and the anatomical limitations of the neck.

Physics and Strategy

Use your sense of touch

Most people rely almost exclusively on sight, and without it we feel extremely limited. But your sense of touch should complement your sight. Furthermore, there are times when you will have to rely exclusively on touch. You should therefore take time to develop this important sense. In a stand-up fight, the most obvious time to use touch is in close quarters, when you are shoulder to shoulder with your opponent. If you feel movement on the right side of his body, he will most likely throw a strike with his right hand. This is because the movement of a strike originates in the body. If he is pressing against you, you can set him up by pressing back and then step to the side and counter-strike. He will now lose balance and fall forward and into your technique. You have added the momentum of your strike to his forward movement, with the result of an increase in power. In essence, you are using your opponent's momentum against him. Strategically, by relying on touch, you will sense what type of technique is coming, and can therefore prepare your defense and counter in advance. Surprising your opponent with a strike when he is off balance will also affect him mentally.

When working at very close range, you can often feel any small move in your opponent's body, and strike the moment he creates a gap.

Because of the close proximity to your opponent in a grappling situation, there are many times when your eyesight will be limited. By relying on your sense of touch, you can still be successful at finding and working a variety of joint locks. I recommend training blindfolded every now and then. This is easier done in grappling than in stand-up fighting, because it is assumed that you will be in contact with your opponent the whole time. When you can control your opponent without actually seeing what you are doing, your awareness will be heightened.

30 Fighting Science

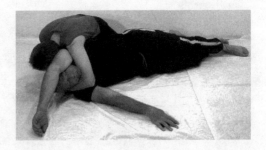

Both fighters are very close and have their heads turned in opposite directions. Who wins this match may be determined by who is able to control his opponent through feel.

In grappling, your sense of touch will be applied mostly through leverage. Presses are a little more difficult, because they require precision against a small target (jaw line, pressure point on arm, etc), which may be difficult to find without sight. Strategically, touch allows you many opportunities. You are not helpless, even if your opponent blindfolds you. When your opponent is behind you, you will still know where his weapons are.

Use several weapons together

Using several weapons together is not the same as throwing a kick and a punch at the same time. You must still ensure that you have full control of balance and can utilize proper body mechanics. In a stand-up fight, this is accomplished by throwing your strikes in combinations. Singular strikes take a lot of effort, because you must overcome inertia with each strike. Combinations allow you to build speed and overwhelm your opponent. Because it is not possible to block all targets at the same time, this gives you many open targets.

Combinations allow you to preserve your momentum and overwhelm your opponent, until you can land a good finishing blow.

Physics and Strategy   31

In a ground fight, it is a little easier to use several weapons at exactly the same time without losing balance or power. For example, with your opponent in the prone position (on his stomach) with you on his back, lock your opponent's legs with your legs and execute a figure-four choke from behind. You will now spread your weight over your opponent and increase your stability, making it difficult for him to throw you off. Strategically, his mind and body focus will split because he will experience pain and defeat at two different points simultaneously.

Martina locks her opponent's leg with her own leg, while simultaneously applying a lock to her opponent's neck.

Use shock value to split your opponent's focus

The purpose of shock value is to get a reaction or to instill fear. In a stand-up fight, this can be accomplished by attacking a vulnerable target repeated times. The nerves on the outside portion of your opponent's thigh is such a target. By flicking a quick round house kick against your opponent's thigh, he will soon become protective of this leg and reluctant to place it within reach. In physics, shock value has very little application, with the strategic advantage being the most valuable.

Muay Thai fighters know the value of a well placed thigh kick.

Some grappling matches last in excess of thirty minutes, where neither fighter is moving very much. But inactivity doesn't win a fight. If you are the lighter fighter, you are likely to find yourself on your back with your opponent straddling you. You will now expend much energy struggling against his weight, but without accomplishing much. Strategically, you can use shock value to get an immediate reaction, which will momentarily split your opponent's focus and enable you to break free. By grabbing soft tissue areas (love handles, inside thighs, inside upper arms) and pinching or twisting, you will create an immediate reaction through pain, even though it will have no lasting effect. You must now take advantage of this reaction and move to a better position.

If you had access to your opponent's face, digging your fingers into his eyes would have a most shocking effect.

Rely on superior defense

It has been said that offense is the best defense. I would like to take that a step further and say that if you can make your offensive and defensive move the same, you will consistently be one step ahead of your opponent. Part of this was demonstrated in the first example on using your elbow to block your opponent's kick. In a stand-up fight, you should also look at using movement and positioning in combination with offense. For example, side-stepping a punch and simultaneously throwing a round house kick will not only accomplish defense, it will have the effect of your opponent walking into your kick. According to physics, this uses the principles of mass and momentum. Strategically, you are cutting the time it takes to accomplish your goal.

Martina places herself in a superior position toward her opponent's back, from which she can counter with a palm strike to the head.

In grappling, too, superior defense relies on offense. We have already talked about how a grappling match often goes into a stalemate, where not much seems to happen. It is possible for a small person to keep a bigger opponent from taking him to submission. This can be done by covering all vulnerable targets, by not presenting him with fingers or wrists, and by not exposing your neck. However, this will only win the fight if your opponent tires much faster than you and is unable to keep the pressure on. In order to reverse the superiority of the fight, you must look for weaknesses in your opponent's defense. Focus away from the area of attack. For example, if he is trying to choke you, rather than grabbing his arms or hands to prevent the choke, try grabbing his chin to get control of his head. Remember, the fighter who controls the head, most often controls the fight. By grabbing the chin, you can raise your opponent's center of gravity and destroy his balance. Strategically, this works because your opponent's focus will be on his attack; on choking you, and not on defending against a counter-attack to his head.

Before your opponent can mount an effective attack, use a defensive throw to unbalance him.

Rely on speed and surprise

It is usually better to act than to react. Whoever initiates action determines the pace of the fight. In a stand-up fight, this means that you must consistently be a step faster than your opponent. He will now be forced to defend against your techniques, rather than you defending against his. Be explosive. Explosiveness comes from rapid acceleration in a short distance. The faster your initial move, the more explosive your technique. Work with different speeds to create surprise reactions in your opponent.

Explosiveness is also reflected in how you carry yourself, as demonstrated by Keith Livingston.

On the ground, explosiveness can be used to reverse positions, or to keep your opponent from gaining a control hold. If you are in the supine position (on your back) with your opponent straddling you, raise your hips explosively and without warning in an attempt to throw him off. It is now imperative that you continue with explosiveness to reverse positions. The problem with working with one speed only is that your opponent will feel any movement in your body and counteract it. Because you give him prior warning, he can apply effective defense, even if you are the one initiating the move.

When held down, your next technique must happen with speed and surprise.

Physics and Strategy 35

Use the inferior position as a strength

Some positions are less desirable than others, because they are difficult to fight from, and because they mentally give you a feeling of inferiority. In a stand-up fight, the inferior position is usually with your back against a wall or the ropes of the ring. By being aware of your position, you can use strategy to reverse it. When you get within one foot of the ropes, start side-stepping the attack. Or if your opponent is pushing you back with his shoulder, start circling with your back away from the wall or ropes and toward the center of the ring. A natural tendency is to automatically mirror your opponent's moves. This can be used strategically to circle your opponent into the inferior position on the ropes. You are now using the principle of non-resistance, and of allowing the momentum to continue. This takes less effort than if you were to pressure back in a straight line. What appears to be the inferior position can now be reversed and turned into the superior position.

In grappling, a wrist lock can be applied in an outside motion (away from your centerline) or in an inside motion (toward your centerline). The drawback of a wrist lock is that it requires fine motor skills to apply. It is therefore not practical in a chaotic situation. A wrist lock should usually follow, rather than precede, some other controlling technique. If you attempt a wrist lock without controlling the rest of your opponent's body, he may try to wrestle out of it. When switching from one wrist lock to another, you can again use the law of non-resistance. When your opponent attempts to wrestle out of a wrist lock, he is likely to go against the direction of the technique. When you attempt to apply an outside wrist lock, he can get out of it by rotating his arm to the inside (toward the centerline). But this also allows you to transition to the inside wrist lock. In effect, if you let your opponent, he will position himself in the inside wrist lock for you. Strategically, he will have accomplished nothing; you still have him in a wrist lock.

When Keith grabs Martina's arm in an attempt to control her, she goes with the motion of the technique underneath the arm and transitions smoothly into a joint lock.

Use logical sequencing of techniques

We have already discussed how there are many strategic benefits that can be attained by working techniques in combinations rather than focusing on single strikes. This section will deal with logical sequencing of techniques. In stand-up fighting, a jab will logically not follow an uppercut in close quarters. Why? The uppercut is a short range technique, while the jab is a long range technique. If the jab is thrown from close range, it will be smothered because you are using the wrong distance for the technique. You will be unable to build momentum for power. Strategically, the jab is not a finishing strike, but a set-up strike, and should therefore precede a power technique, such as the uppercut. If your opponent moves back and creates distance, it would be better to follow with a rear cross or an overhand strike (which are both finishing strikes).

The bottom line is that if your techniques are logical, they will be thrown with less effort and more speed. Through speed comes power. Logical sequencing of techniques will keep your fighting range optimal. It is possible to throw an uppercut following a side kick, but because the side kick is likely to knock your opponent back, and the uppercut is a close range technique, if you attempt this combination, you will need to utilize more and bigger movements to close distance. Aside from telegraphing the technique, this is also uneconomical.

In grappling, too, logical sequencing of techniques should be considered. For example, if you execute a rear choke on an opponent on his stomach, it would not make sense to follow this technique with a finger lock. A better combination would be to work a series of techniques that eventually lead into the rear choke, and to finish with the choke.

Indomitable spirit

One last thing that ought to be mentioned is the martial spirit. When I was studying karate, we were told that one of the characteristics of a superior fighter is *indomitable spirit*. The instructor explained that this means to have a spirit that "can't be dominated." When I became an instructor myself, I did some research into the word origin and found that it stems from the word diamond, which, in turn, means untamed. A diamond is the hardest material we know of, and the only way to cut a diamond is with another diamond. In essence, indomitable spirit means a spirit that can't be broken.

The reason a diamond is so strong is because the molecular pattern is so complex. I once observed a board breaking demonstration where a young kid was going to break a one-inch board with his fist. When he struck the board, it didn't break. The young martial artist grabbed the board from the holder, took one quick look at it, and turned it one quarter of a turn. When he struck again, the board broke. This youngster knew something about the building material of the board, and where the point of strength and weakness were. If you attempt to break against the grain, it will take more force than if you try to break with the grain. Likewise, if the grain is wavy or has a knot in it, a greater force is required to break the board. It can therefore be said that the more complex the grain (the pattern), the more force is required to break it.

Because of the complex molecular structure of the diamond, you wouldn't dream of trying to break one with your fist. Much of the value of the diamond is also derived from it being multi-faceted. In martial arts, you can be multi-faceted by cross-training in different arts. On a grander martial scale, this can be applied to whole armies. The general who leads an army of a thousand men, whose only quality is muscular strength, will soon be defeated by the opposing army who has a mixture of men with qualities such as strength, speed, intelligence, etc.

> **"The brave can fight, the careful can guard, the intelligent can communicate."**
> — Sun Tzu, The Art Of War

The more you diversify your skill, the more likely you are to succeed in battle. I am not implying that you should be a "jack of all trades," but that you should educate yourself on how to use your skill in many different situations. If all you know is how to strike and kick, and you end up on the ground, the battle may well be over. The opposite is also true. Today, when cross-training is so popular, we see a lot of grapplers studying Muay Thai and kick-boxing, and vice versa. In a nutshell: don't place all your eggs in one basket . . . and hide an ace in your back pocket.

With this said, follow me on the *power trip*, and I will show you the easy way how F=ma.

Center Of Gravity
Balance, How Important Is It?

When looking back, I consider my karate training much more ritualized than the kick-boxing training I later received. Karate students were expected to do things a certain way and, if deviating from this pattern, we might suddenly find ourselves on the floor without really knowing how we got there. Certain rituals had to be adhered to: bowing when stepping into the Dojo, lining up according to our rank, and keeping a strong focus on the instructor and the techniques taught. After warm-up and stretching, we often did drills in the basic punches and kicks. This was always done from a ***horse stance***, so called because of its similarity to a person riding on a horse.

We were told to keep our feet about shoulder width and a half apart, with our knees slightly bent. The instructor would call out the punches and we would follow with a loud *kiai*. If the classes were large, there were often one or more assistant instructors, who would walk between the rows of students. Every so often, I would see, from the corner of my eye, a student in an adjacent row tumble to the floor. The instructor had swept the student's foot, taking him off balance. Every time the instructor neared me, I feared I would get swept. This was not a malicious move on the part of the instructor, but simply an attempt to make us understand the importance of balance and correct posture.

Those who have studied martial arts for a long time know intuitively how to maintain balance, how to strike with power, and how to use defense and footwork effectively. But if you can add to that an in-depth understanding of the laws of physics, you can speed up the learning process considerably. You will know from the beginning if a technique is mechanically correct and why. This may save you years of work by eliminating trial and error. All the principles of physics are important and can be combined to strengthen your fighting tactics. However, if I had to place more emphasis on one principle, it would be on ***balance***. If you lack balance, it doesn't matter how powerful, fast, mechanically correct, or skilled you are, none of your techniques can succeed. Because balance is literally your foundation, we will talk about it first.

Balance and stance

In addition to strengthening the legs, the horse stance serves the purpose of stability. In general, when we talk about physical stability, we mean an object or person who possesses balance and will not tip over easily. Wrestlers and grapplers, for example, need a great deal of balance to counteract their opponent's attempts to throw them to the ground. Stand-up fighters also need balance to deliver a strike or kick with power.

Successful fighting depends on gaining a positional advantage, and a fighter seldom stays in one place. It is now necessary to learn all over again how to "walk." Moving from one spot to another is not as simple as taking a step forward. Many factors need to be considered; among those superior positioning and balance. This has led to the development of the ***Basic Movement Theory***, which states that *whenever moving, you should step with the foot closest to the direction of travel first.* For example, to move forward, you should step with your lead foot first and readjust the width of your stance with your rear foot. To move backward, you should step with your rear foot first and readjust the width of your stance with your lead foot. The primary reason for the development of the Basic Movement Theory is to keep a fighter from crossing his feet.

Which fighter appears to be in the more stable stance? Why is the horse stance more stable than a stance in which your feet are close together or crossed?

Center of gravity

A few years later when I became an instructor myself, I loved the push-up routine. Schools of self-defense often advertise that "you don't need size and strength to defend yourself effectively against a bigger opponent." Being a skeptic, I insisted upon my students developing strength. We would do ten push-ups, and then I would ask them to hold the down position without touching the floor with any body part other than the hands and feet.

Center of Gravity 41

When researching push-ups in more depth, I found that the "girl push-up" (on your knees) is nearly useless for developing upper body strength. This is because you lift such a small percentage of your body weight. I also found that, in the regular push-up, men lift slightly more of their body weight than do women. I would tell my students this is because women have bigger butts (because I am a woman, I felt it was safe to say this). In relation to overall body structure, the average man is more top heavy than the average woman. But is the bigger butt necessarily such a bad thing? It all depends on what you are trying to achieve. *Whether the glass is half-full or half-empty depends on what's in the glass, and how desirable it is to the person who has to drink it.* I found that being bottom heavy is advantageous to the martial artist.

Center of gravity (or center of mass) is the point in any object around which all of its weight is equally distributed. We can also think of center of gravity as the balance point. An object of uniform shape and weight has the balance point in the center. An object of non-uniform shape and weight has the balance point toward the heavier end. Assuming that the weight is equally distributed throughout the following objects, where will you find each object's center of gravity?

The center of gravity is always found toward the heavier end of the object. On a horseshoe shaped object, the center of gravity is in the open space toward the closed end of the arc.

A pyramid has a low center of gravity and is very stable because of its wide base. The wider the base, and the more weight that is concentrated in the base, the more stable the object is. For maximum stability, you would therefore keep a stance that is as low and wide as possible. There is, however, a trade-off. If your stance gets too wide, you will sacrifice mobility for stability. If fighting relied on stability only, you would study arts that employ very low and wide stances. But since force is a combination of *mass* and *acceleration*, which in turn branch off into many other components, fighting becomes a "give and take" situation, where you might need to make a sacrifice in one area in order to gain an advantage in another.

Take a look a these two pyramids. Which one is more stable?

In both cases, the base is wider than the top. But in relation to its own height, the pyramid on the left has a wider base than the pyramid on the right. Let's look at what exactly it is that makes the pyramid on the left the most stable of the two.

Stability depends on two factors:

• How high you need to raise the center of gravity in order to make the object top heavy. The wider the base, the higher the center of gravity must be raised before the object will tip over.

• If you drop a line straight down from the center of gravity of an object of any shape, and it falls within the foundation of that object, the object is stable. If it falls outside of the foundation, the object is unstable.

I'm sure you can see that quite an effort would be needed to tip the left pyramid on its side. In the right pyramid, the center of gravity is already quite high, so this pyramid will be easier to tip over.

In addition to being more stable, the pyramid on the left has its weight focused over a larger surface area than the pyramid on the right, resulting in less *force per square inch*. This principle can be applied to finger push-ups. Because the fingers are inherently weak, finger push-ups take a great deal of strength. The more you can spread your weight, however, the easier the push-up will be. When starting to train in finger push-ups, keep your hands and feet wide apart. This will spread your weight over a large surface area with less force per square inch.

A handstand takes both strength and balance, but would be even more difficult if your hands were close together. In the picture below, note how the body is pyramid shaped for greater stability.

Handstand with wide base for balance.

If you are a grappler, less effort is needed to pin your opponent to the ground if you stay as low as possible. Spreading your weight and staying low (wide base, low center of gravity) will make it difficult for your opponent to throw you off, even if he has the size and strength advantage. Standing on your opponent or keeping an upright posture (narrow base, high center of gravity) will make it easier for him to throw you off.

In both cases below, the fighter on the bottom is in the prone position face down. Would you say that she has a good chance of winning this fight? Well, much of it depends on how stable the fighter on top is. If you were the fighter on the bottom, who would you rather deal with: a bottom heavy or a top heavy opponent?

44   Fighting Science

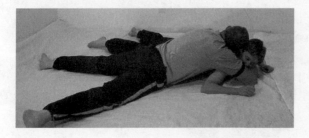

Wide base, low center of gravity (stable)

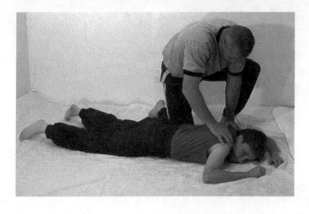

Narrower base, higher center of gravity (not as stable)

The second factor that determines stability is whether a line dropped straight down from the center of gravity falls within the foundation of the object. This concept becomes especially important whenever we throw a kick with our *lead leg*, as we shall soon see.

Line dropped straight down from center of gravity falls *outside* of foundation.

Of the two objects above, the one on the left is more stable than the one on the right, not only because it has a wider base, but also because a line dropped straight down from the center of gravity will fall within the base of the object. The object on the right, however, is *lopsided* and will tip over.

Center of Gravity   45

This gymnast is in a balanced state because her center of gravity is directly above the narrow base of her hands. The karate practitioner performing a side kick is also in a balanced state because his center of gravity falls directly above the narrow base of his foot. If the karate practitioner's upper body wasn't angled backwards in the opposite direction of the kick, he would be unable to maintain balance. Try it!

The reverse handstand seen in the pictures below is an exercise in balance more than in strength. In order to perform a handstand at a diagonal angle, it helps if the hands are reversed. This is because the hands (the foundation) are closer to the head than to the feet. If the fingers were pointing toward the head, the foundation would shift horizontally closer to the head, the center of gravity would not fall above the foundation, and maintaining balance at this angle would be difficult. Note also, as the legs are lowered, the upper body must move slightly forward as a counter-weight.

Even though the pyramid shape may be the most stable, it does not mean that a person standing on one foot will lose balance; only that it is easier to unbalance him than if he were in a wide horse stance. The best time to unbalance your opponent may therefore be when he is on one foot in the process of kicking or stepping. When the foundation is narrow, the center of gravity needs to shift a lesser distance to make you unstable, than when the foundation is wide. If you are in a horse stance, your center of gravity must shift as much as two feet before you become unstable. If you are on one foot, however, your center of gravity must shift only enough to fall outside of the width of your foot, which may be an inch only. If you are fighting against a high kicker, a kick or sweep to his supporting leg may be a good technique to get him on the ground.

Side-stepping a kick, as demonstrated by Cindy Varnell Winlaw, and simultaneously attacking your opponent's supporting leg, is a good technique for unbalancing her.

Although a wide stance is more stable than a narrow stance, there is an interesting twist to this. If you sweep one of your opponent's feet when his stance is very wide, a lot of upper body movement is required to shift his center of gravity above his supporting foot. Balance will now become a *time and reaction* issue, and your opponent may actually lose balance easier than if you sweep when his stance is narrow.

It is possible to be stable in one direction, yet unstable in another at the same time. The wider your base is in the direction of the applied force, the more stable you will be. Thus, a horse stance is unstable against an attack from the front, yet stable against an attack from the side.

The power of the lead leg kick

When I was studying karate, I used the lead leg front kick a lot, because it was fast and easy to throw, and could score a point on an opponent with ease. When I started in kick-boxing several years later, my instructor told me that my lead leg front kick was *lousy* and totally lacked power, and that his grandma, who was 85 years old and had had hip surgery, had a stronger kick than mine.

I went home and practiced on the heavy bag, and discovered that the rear leg front kick was slower but more powerful than the lead. I also found that my lead foot was often slightly short of reaching the target, or that the foot was sliding against the heavy bag instead of penetrating it. Because competition kick-boxing relies on powerful strikes and kicks more so than competition in karate, where the fighter relies on scoring points rather than knocking his opponent out, the question was whether I should abandon the lead leg kick and throw all of my front kicks with the rear leg. But because the lead leg is so quick, I didn't want to give it up, and so I began to research the reason behind the lesser power of the lead leg.

In order to maintain balance when kicking, you must do one of the following:

- Center your upper body (center of gravity) above your supporting foot (base).

- Center your supporting foot (base) below your upper body (center of gravity).

Unless you do one of the above, your center of gravity will not be above your foundation, and you will lose balance. Why is this important to power, and especially so when throwing a kick with your lead leg? Sometimes, the power of the lead leg is a flexibility issue but, assuming that you have good flexibility in both legs, the lead leg is less powerful because:

- The distance between the lead leg and the target is shorter than the distance between the rear leg and the target. The shorter the distance, the less time you have to build *momentum* for power. We will talk more about momentum in Chapter 3.

- The center of gravity (your upper body) must be above your supporting foot (in this case, your rear foot). This moves your upper body to the rear, making the lead leg slightly out of reach.

This fighter's center of gravity is located in his upper body, aft of the lead foot and forward of the rear.

When the fighter raises his leg, he also centers his upper body above his supporting foot.

When the fighter extends his leg to kick, his upper body moves an additional few inches to the rear for balance.

Because the center of gravity and the foundation must be vertically lined up for balance, when your lead foot comes off the floor, you must also move your upper body slightly to the rear until it is centered above your rear foot. The drawback is that your upper body will move in the *opposite direction* of the kick, causing opposing movement and splitting the power into two directions (most of it forward, but some of it back). The remedy is to move your *rear foot forward* and underneath your upper body, instead of moving your upper body back and above your rear foot. This can be accomplished by taking a small step forward with your rear foot prior to kicking.

A step forward with your rear foot, until it is centered underneath your upper body, helps you achieve balance when throwing a lead kick. Note how the weight is on the rear leg.

So, what's the trade-off? Well, whenever you step prior to kicking, you will add movement and the time it takes to step, and will therefore risk telegraphing the kick. Stepping with your rear foot also narrows your stance for a short time, which defies the Basic Movement Theory. In order to throw a powerful lead leg kick, you must make a sacrifice in another area: *movement, time, and balance*.

Note: A step prior to kicking with your *rear leg* is not required, as your upper body will automatically move forward above your supporting lead foot. The forward movement of your upper body will also increase the power of the kick.

Balance in the hard and soft styles

How balance is maintained is also reflected in the many different styles of karate. Karate is divided into "hard" and "soft" styles. Somebody not educated on the subject might think that hard and soft refer to the effectiveness of the style, or how painful the techniques might be. That is not the case. The ***hard styles*** (in general the Japanese styles) utilize wide stances and linear moves, whereas the ***soft styles*** (in general the Chinese styles) utilize more flexible stances and circular moves. There is also a wide variety of styles that fall in between these extremes.

Practitioners of Shotokan karate, one of the hardest of the hard styles, rely on wide stances to maintain balance. The soft style of Kung-Fu, on the other hand, utilizes the movement of the body. This is similar in concept to the way a chicken or pigeon maintains balance: the head must move forward or back each time a step is taken. The strikes are also circular in nature.

The hard styles rely on the "one-strike" concept, with maximum power in the linear direction. These linear strikes usually start from long range, allowing the practitioner to build momentum through forward movement. The linear strike is economical in time (the shortest distance between two points is a straight line) and usually focuses on the opponent's centerline, where some of the most vulnerable targets are located. Hard strikes are great for catching an aggressive opponent off balance, especially if he is in the process of throwing a kick.

The linear strike can be likened to a block (as opposed to a parry) that meets power with power. When your opponent is stopped in his tracks, he will experience a jarring effect that will set him back, push him off balance, and force him to regroup.

The drawback of the hard style "one-strike" concept is that all energy is focused into that one strike, and if the strike fails to do the intended damage, it will take both energy and speed to reposition for a follow-up technique. The soft styles, on the other hand, give the practitioner a bit more versatility. Soft arts rely on circular motion with strikes in combinations, where the "finishing" strike builds speed for power through one or more "set-up" strikes. This circular motion allows the practitioner to build tremendous momentum, and to work at close range from a superior position around his opponent's back. Multiple strikes in combinations also tend to split your opponent's focus and cause **sensory overload**.

In principle, a soft strike can be likened to a parry. The idea is to keep your opponent's motion going, thereby unbalancing him and drawing him into a follow-up strike. Soft styles are yielding, allowing you to use your opponent's own energy against him.

We, as fighters, have a natural tendency to focus our attacks straight ahead, but there are many advantages to working the angled attacks of the soft styles. When going directly against the force, balance can only be manipulated if your opponent already has a narrow base. This is why the sideways fighting stance is so beneficial in the frontal direction where most attacks are aimed; you can brace yourself with your rear leg. The soft styles, on the other hand, allow you superiority to the side or back of your opponent's body, allowing you to attack at an angle away from his focused power.

Balance manipulation in throws

Successful throws rely on your ability to manipulate your opponent's balance. When two people are tied together, as is the case when one person is throwing another, an interesting thing occurs: *In regards to balance, the two fighters will begin to act as one.* Physics does not know that the two fighters are separate entities. With this in mind, ask yourself what the objective of the throw is and where the center of gravity is located.

The objective of a throw is to get your opponent on the ground, but without going down with him. If you grab your opponent and hoist him over your shoulder in preparation for a throw, *your* feet are the only foundation, but *both* your bodies joined together will determine where the center of gravity is located.

Center of Gravity   51

In a hip throw, the thrower's two feet act as the foundation. If the combined center of gravity for these two fighters were to fall outside of the foundation, the throw would fail, and both fighters would lose balance.

Before a successful throw can come to completion, it requires a balanced state of the two bodies. Note, in the picture below, how the combined center of gravity is above the very small foundation of the thrower's left foot. Because the person initiating the throw has maintained balance between the two bodies, the next step, the throw itself, will have a potentially successful outcome.

Balanced state at the initiation of a throw.

The final phase of throwing relies on the principle of making only your opponent's center of gravity fall outside of the foundation, while your own center of gravity remains above the foundation. To clarify, in the next picture, the foundation is the distance between the thrower's two feet. However, the center of gravity of the person being thrown falls outside of the thrower's foundation. Because the two bodies act as a unit, when the thrower begins to feel the effects of the throw, he must allow his opponent fall, or the two bodies will continue to act as one, their combined center of gravity will not be above the foundation, and the thrower will go down with his opponent. Note also how the thrower's upper body is angled forward as a counter-weight in the opposite direction of his opponent's feet.

52   Fighting Science

Successful throw with full control of balance.

Correct mechanics, and full understanding of center of gravity, enables a lighter and weaker person to throw a heavier and stronger opponent. Once your opponent starts going over your hip, his center of gravity will fall outside of the foundation and will begin to act to your advantage. As long as your body mechanics are correct with your own balance maintained, the rest of the throw will happen almost automatically. A 100 pound girl, who has received the appropriate training, can with ease throw a 200 pound man in this way.

Balance manipulation in joint locks and takedowns

One of the easiest ways for a smaller person to defeat a bigger adversary is by attacking the inherently weak areas on his body. It doesn't matter how big or muscular your opponent is, he cannot flex his eyeballs, ear drums, or testicles. But achieving a position from where you can easily use a strike against these targets is a different matter. If you can get your opponent on the ground without going down with him, you have a distinct advantage. You have now bought time which may allow you to escape, while your opponent must expend energy defeating the forces of *gravity* and *inertia* to get back on his feet. Balance manipulation is one of the easiest ways to get your opponent on the ground. In addition, since the neck is an inherently weak area with a high center of gravity, if you can use the neck to shift your opponent's balance, not much physical strength is needed to take a bigger adversary down.

Keith Livingston uses neck manipulation in conjunction with a joint lock to shift his opponent's center of gravity.

Center of Gravity

The closer your technique is to your own center of mass, the more strength you will appear to have. Which is easiest: To lift a 100 pound box with your arms extended and the weight away from your body, or with your arms bent and the weight as close to your body as possible? The latter is true, because keeping the weight close to your body allows you to use the strength of your entire body to produce the force, and not just the muscular strength in your arms. The principle of physics that allows you to do this is called *torque*. We will talk about torque in greater detail in Chapter 5.

The same concept is true for takedowns. Keeping the technique close to your center of mass allows you to use your body weight and not your muscular strength. Most takedowns also rely on circular motion. The farther a technique is from the axis of rotation, the more movement is needed. Energy can be conserved by keeping the technique close to the center of the circle.

In the pictures below, the martial artist utilizes two principles of physics: First, he pulls his opponent forward and off balance, relying on his opponent's forward momentum. Second, he executes a joint lock against the wrist followed by a takedown, using very little strength because the technique is so close to his own center of mass.

Keith pulls his opponent off balance and close to his own center of mass, where he executes a takedown through a joint lock.

Because our joints work best in one direction only, going against the natural movement of your opponent's joints gives you a clear advantage. Let's say that you approach your opponent from the front, grab his hair, and pull until he is stooped forward. Will this result in a takedown? Not likely, because pulling your opponent's head toward you allows his hips to move to the rear. Unless the move is done violently, his center of gravity will have time to shift and stay above his foundation. The takedown will therefore fail. This is also why you can bend forward and touch your toes (providing that you have the flexibility) without losing balance. However, if you were to push your opponent's upper body back rather than forward, it would be more difficult for him to retain balance. This is because the body is not naturally designed to bend backwards very far. To achieve a takedown, you only need to push your opponent's upper body back enough to shift the center of gravity so that it is no longer above his feet. Providing that you don't allow him to step and regain balance, the takedown is now not only possible; it is certain. In fact, once the center of gravity has shifted away from the foundation, you can let go entirely and watch your opponent fall.

Martina places one hand on her opponent's chin, and the other on the back of his head. She pushes his upper body back until the center of gravity has shifted to the rear.

The effectiveness of a takedown can be increased by attacking two different points on your opponent's body simultaneously. This keeps him from using the natural movement in his joints to escape the takedown. The same principle applies regardless of which part of the body you attack. For example, in an upper body takedown, you may push with one hand on your opponent's jaw to tilt his head back, and push with your other hand against the small of his back to move his hips forward. In a lower body takedown, you may place one hand around your opponent's heel and pull, and the other above his knee and push. By attacking two points of different height, you can isolate the joints and increase the effectiveness of the takedown.

Center of Gravity 55

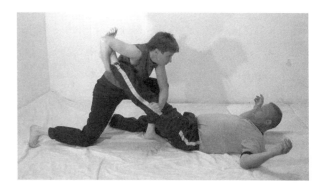

Martina ducks her opponent's strike and attacks two points of balance: behind the ankle and above the knee. Had she attacked one point of balance only (the ankle), her opponent would have retained flexibility in the leg, and the takedown may not have worked.

We have all heard the expression "the eye of the storm." In a rotating system, the highest speed occurs away from the center. The higher the speed, the more chaotic the takedown, and the more difficult it will be for your opponent to retain balance. Think of yourself as the eye of the storm, with your opponent whirling around you. Being the center of the rotating system, where it is relatively calm, enables a smaller person to unbalance a bigger adversary without too much of a struggle. You are now relying on your opponent's momentum to throw him off balance. We will talk more about momentum and *rotational speed* in Chapter 5.

When Martina is threatened by a much bigger opponent, she steps forward and grabs him around the neck, simultaneously starting the rotation. Note the two points of balance: the neck and the hip. Because of the circular momentum, not much effort is needed to shift the center of gravity until the opponent goes down.

Keith uses a joint lock in conjunction with circular momentum. Because the technique goes against the natural movement of the joint, attacking two points simultaneously is less crucial. In fact, the elbow will act as the second point of balance in conjunction with the wrist lock.

Because loss of balance is such a chaotic feeling, unbalancing an opponent gives you a strategical advantage, as well as a safety advantage. When your opponent is about to lose balance, his mental focus will be on regaining balance, and whatever technique he had in mind cannot be completed. In addition, getting back up after a takedown is both time and energy consuming. However, it is worth mentioning what can go wrong when attempting a takedown. First, you must act dynamically. Rely on the moment of surprise. Especially in body takedowns, your positions are often equal initially, and if you don't act quickly, the stronger fighter will win. A distracting technique can be used to split your opponent's focus, followed by an immediate attempt to unbalance him. Once you have initiated the takedown, your energy must be focused in the proper direction. Thus, if you want to take your opponent down by pushing his upper body back, your focus must be down and not straight back. Likewise, if you rely on circular momentum, but don't apply downward movement simultaneously, you are likely to pull your opponent around in a circle rather than taking him down. **Vectors**, and how to avoid splitting the power, will be discussed in Chapter 4.

Does the same principle apply on the ground? Sure it does! If your opponent is on his hands and knees, how would you unbalance him? Remember, the principle is shifting the center of gravity away from the foundation. This can be done by pulling one of his hands out from under him, while pushing against his shoulder. The two points of balance are the hand and the shoulder.

An opponent on all four can be unbalanced by pulling one hand out from under him.

Another option is to grab and turn your opponent's head until his body must follow. The two points of balance are the chin and back of the head.

If you are on your back and your opponent is straddling you, the principle remains the same. Try shifting the center of gravity with a joint lock or with head manipulation. The reason a joint lock is often effective is because it causes extreme pain, so it is more difficult for your opponent to focus on retaining a balanced state between his center of gravity and his foundation.

An opponent who is straddling you can be unbalanced by shifting the center of gravity through head manipulation.

Physics, in the martial arts, is not about which style or technique is better. Forget about that argument. Rather, it is about taking what you have, regardless of style or technique, and using that to achieve your objective. It is not about whether you are standing or on the ground, or whether you use a sweep, a throw, or a joint lock to unbalance your opponent. It is about shifting the center of gravity--*always*. One principle, many applications. Physics is only boring or difficult when you don't see the application. Physics gives you options. Making physics your ally allows you to take advantage of whatever position you happen to find yourself in. That's why understanding physics makes you great!

Summary and review

Balance may be the most fundamental concept of the fighting arts. When your opponent is off balance, his power, speed, stamina, and skill go out the window, and the most vicious adversary becomes an easy take. Many techniques can be used to steal a person's balance (sweeping a leg, tackling and throwing, applying a joint lock, etc.), but they rely on one principle only: the location of the center of gravity--the point in your opponent's body where all his weight seems to be focused. If the center of gravity is not above his foundation, he will go down, regardless of which technique you use, or what kind of defense or counter-attack he attempts. Understanding this one principle leaves you free to choose whatever technique is most appropriate at the time of the encounter.

The principle of balance states that:

the C.G. (center of gravity) must be above the foundation

Your C.G. is the balance point in your body. Thus, if you have muscular shoulders, you will have a higher C.G. than if you have a big belly. Although muscular shoulders may be desirable, a low C.G. is more stable than a high C.G. So, next time you admire yourself in the mirror, remember that you must trade your muscular good looks for stability.

The C.G. is usually located in the lower part of your upper body, but you can raise or lower your C.G. depending on how you distribute your weight. This means that a fighter in a low stance is more stable than a fighter in a high stance, who may be top heavy. The wider your foundation, the more the C.G. must shift in order to unbalance you.

If you bend forward or back, or lean to one side, your C.G. will shift. If you lean too much, you will lose balance. Whenever your C.G. does not fall within the area of your foundation, you will lose balance. No exceptions! Test this principle by standing in a horse stance with your weight equally divided between your feet. Now, raise one foot off the floor. Unless you simultaneously move your upper body above your supporting foot, you will fall.

We will now look at how to maintain balance when attacking and when under attack. We will also look at how to take advantage of an opponent who is in an unbalanced, or nearly unbalanced, state.

Balance in stances

Stance involves more than *standing*. Your stance is your foundation. If you practice a stand-up art, it is the distance between your feet. If you are on one foot, your foundation is the area of that one foot. If you practice ground combat, your foundation is the area between the body parts that touch the ground.

Wide stances, and why they are stable:

1. A wide stance is more stable than a narrow stance. This is because your foundation covers a larger area (the area between your feet).

2. There are exceptions to the stability of the wide stance. A very wide stance may become unstable easier than a narrow stance. This is because you must move your upper body a longer distance to place the C.G. above your supporting foot, should the balance of one foot be taken.

3. A stance that is too wide may not be practical for fighting, because it is not very mobile. In addition, it does not allow you to take advantage of position or body mechanics to effect a powerful strike.

4. A wide stance makes it more difficult to place your body mass behind the strike for power. Pivoting, yet keeping your balance centered, may also be more difficult from a wide stance.

Narrow stances, and when they will benefit you:

1. Narrow or crossed stances are unstable when taking a strike, because the C.G. does not need to be shifted as far in order to fall outside of the foundation.

2. Although the narrow stance is less stable than the wide stance, there are times when a narrow stance will benefit you. An example would be as a propellant to move you forward, as when throwing the crossover side kick. The momentum of the crossover step may outweigh the benefits of a wider stance.

The crossover step is a potentially unstable maneuver that is often used as a propellant when closing distance with a kick.

3. You should also look at how to take advantage of your opponent when his stance is narrow. One of the best times to counter-strike may be when he is on one foot in the process of kicking. His base is now narrow and, although he is not off balance initially, it will be easy to shift his C.G. so that it falls outside of the foundation.

High and low stances:

1. A low stance is generally more stable but less mobile than a high stance. The low stance is advantageous when you don't desire to move, as when in a clinch.

When in a clinch, keep a low and wide stance for balance.

2. Be careful not to lean on your opponent when in a clinch, as this may make you top heavy. If he takes a step back now, you will fall forward and into his counter-strike.

62 Fighting Science

Balance when striking

Just like stance involves more than standing, striking involves more than extending your arm. Any time you extend an arm or a leg, you will affect the center of gravity, and must therefore make an adjustment in your body.

The importance of bending the knees:

1. Your knees should be slightly bent for stability when striking. If you straighten your leg or allow one foot to come off the floor, you will also raise or shift your C.G. so that it is no longer above the foundation, or so that it is too close to the limits of the foundation.

If you allow your foot to come off the floor when punching, you risk overextending the center of gravity forward.

2. If you are off balance, your strike will lack power because you can't utilize body momentum maximally. Straightening your knee may also allow your upper body to move forward and starve the strike of optimum distance.

3. Your stance should be relaxed. If your knees are straight, you are more likely to tense. As a result, it will be difficult to make adjustments for any shift in C.G.

Sideways stance vs. horse stance:

1. The drawback of fighting from a sideways stance is that your lead side is more exposed than your far side, which makes this an easy and quick target for your opponent. Be aware of techniques that are intended to unbalance you, like a sweep to your lead leg.

Center of Gravity 63

The toe-to-heel stance allows you balance and flexibility in all directions.

2. A horse stance is unstable from the front, yet stable from the side. In order to gain reasonable stability in all directions, many martial arts employ a "toe-to-heel" stance. An added benefit is that this stance gives you reach with both your lead and rear hand.

3. The most dangerous time to get in a horse stance may be when you are cornered or with your back to the ropes. If your opponent presses you now, you will get back on your heels, unable to step back to brace yourself.

Balance when kicking

The principle of keeping your C.G. above your foundation applies to both striking and kicking. The reason it is more difficult to maintain balance when kicking is because your foundation is narrow (your foundation will never be wider than your supporting foot). This is also why the best time to take your opponent's balance may be when he is in the process of kicking.

Natural combinations:

1. A natural combination allows each punch and kick to flow easily off the previous one. In a natural combination, there are no awkward movements in your body, and you are never at a danger of loss of balance.

2. Natural combinations when kicking generally employ alternating kicks, relying on the momentum from one kick to set the next kick in motion.

3. Kicking multiple times with the same leg requires very good balance, especially if opting not to plant the foot between kicks. Planting the foot between kicks allows you to ricochet off the floor when reversing direction, thus using less muscular effort.

Using the opposite side of your body as a counter-weight to balance:

1. The more rigid you are, the more difficult it will be to maintain balance. The different parts of your body should therefore be allowed to move independently. This enables you to shift your C.G. so that it is always above your foundation.

2. Many kicking arts rely on "pulling" with the opposite side of your body. This allows your hips to follow through for maximum momentum. It also allows you to shift your C.G. so that it is always in a balanced state.

3. Your hands may be lowered or moved to the rear as a counter-weight to balance. The drawback of this is that your head is exposed. It may be better to keep your hands high and bend your knees. You can also crouch to increase balance.

An arm may be lowered or stretched in the direction of the kick to increase balance and momentum.

Center of Gravity 65

Chambering for balance:

1. Chambering your leg prior to extending the kick helps with balance. There is less *inertia* when your foot is close to the axis of rotation, so a chambered kick requires less muscular effort to lift off the floor than a straight leg kick, and can therefore be launched quicker.

Chambering the leg and kicking.

Raising the leg straight and kicking.

2. Pulling your leg back to the fully chambered position before replanting your foot on the floor enables you to shift your C.G. to a stable position which, in turn, enables you to throw successive kicks without loss of balance.

Unbalancing your opponent:

1. Any time you can grab one of your opponent's legs, you can unbalance him by pushing back on that leg. You will now move his C.G. to the rear, without giving him the opportunity to move his foundation.

66   Fighting Science

2. Sweeps or sweep kicks are good for unbalancing your opponent. Sweeping toward your opponent's centerline will narrow his stance with the effect of him going down. Sweeping away from his centerline will widen his stance with the effect of his upper body coming forward.

Sweeping toward centerline.   **Sweeping away from centerline.**

Balance in defense

As much as half of fighting may be about defense. The fighter who can use his defensive techniques offensively (or to launch offense) is ahead of the fighter who thinks about offense and defense separately. The principles of physics can be applied equally to offense and defense. If you are in an unstable state when blocking, the power and momentum of your opponent's strike may unbalance you. Just like offense, defense should therefore take place when you are in a balanced state, preferably when your foundation is wide and you are crouched with a low center of gravity. Much of defense is also about learning how to avoid your opponent's attempts to use these same principles against you. Focus on keeping your own balance when defending, and on unbalancing your opponent when he is defending.

Principles of linear blocks:

1. A linear block meets the strike straight on. Examples of linear blocks are upward block, downward block, elbow block, and shin block. Inward and outward blocks that meet the strike at a perpendicular angle are variations of linear blocks.

Center of Gravity   67

2. When utilizing the upward block, lower your C.G. slightly for balance. Don't raise up on your toes or lean back. Because a strike that comes down on you has a lot of power, don't step to a narrow stance when blocking.

3. When utilizing a downward block, lower your weight for balance. This will also increase your momentum and power.

4. When utilizing the shin block, because you will be on one leg, you must be extra careful not to let your opponent's kick knock you off balance. Try to meet the kick, rather than waiting for it to come to you.

When raising your leg to block with the shin, try to keep your center of gravity above your foundation and slightly forward and into the kick.

5. The inward and outward block might require a slight turn or pivot in your body for evasion. Don't lean back when pivoting, as this may make you top heavy to the rear. Your body should remain upright with the C.G. above the foundation. If your opponent leans back when he is blocking, take advantage of it by sweeping his leg. Because his C.G. has shifted to the rear, he will easily go down.

Inward block without leaning.

If you lean, your opponent may take advantage of your rearward center of gravity.

68   Fighting Science

Principles of redirecting blocks:

1. A redirecting block allows your opponent's strike to continue with a slight change in direction. The parry is a redirecting block.

2. Don't lean when parrying. Leaning shifts your C.G. farther from the center of your foundation. Try to intercept the strike before it requires you to evade.

3. If you step when parrying, step to widen the stance. Don't cross your feet.

Your body should remain upright when parrying (left). If you step with the parry, be careful not to cross your feet or narrow your stance (right).

Principles of evasive movement:

1. Evasive movement allows you to evade an attack all together, without making any contact with your opponent's strike. Slipping, bobbing and weaving, and pivoting are examples of evasive movement. The pivot can be done in conjunction with a block or parry.

2. When pivoting with a step, your weight should shift until it is above your supporting foot. You can also pivot without moving your feet, using a broad shift in weight and moving your upper body only. The knee of your supporting leg must bend slightly to allow for a quick shift in C.G. should your opponent try to unbalance you. Bending the knee also lowers your C.G. for stability.

Center of Gravity   69

3. When slipping, move your head only. Don't lean. This is especially important in the rearward slip. Your weight can shift from foot to foot, but your knees should be bent to prevent leaning. Shifting your weight from foot to foot gives you an automatic slip.

When slipping, keep your knees bent to allow for any shift in center of gravity.

4. When bobbing and weaving, bend your knees and keep your back straight. Bending at the waist with straight legs makes you top heavy forward.

When bobbing and weaving, keep your back straight and stay focused on your opponent.

Your opponent's balance:

1. The best time to attack your opponent's balance may be when he is on one leg. The need to defend against his kick can therefore be taken advantage of by moving to close range inside of his kick, simultaneously attacking his supporting leg.

2. Your opponent's balance can also be taken from long range by blocking his kick with your elbow (which also gives you a high force per square inch), and simultaneously attack his supporting leg.

3. Your opponent's foundation is narrowed whenever he is stepping. Attack when he is in the process of stepping backward in a defensive move.

Attack when opponent is back-pedaling.

4. By using a parry rather than a block, you can unbalance your opponent by allowing his motion to continue.

Opponent's motion continues, but is redirected with the parry. This unbalances him forward.

Your own balance:

1. Your own balance is weak when you are on one leg, as may be the case when defending with a shin block against a kick to the leg. Be prepared to meet the strike and counter.

2. Try to avoid defending with two blocks simultaneously (a shin block and a parry, for example, or a shin block and a bob and weave). Balance suffers when there are opposing movements in body mechanics.

3. Be careful not to block a strike when your feet are crossed or your stance is narrow. Even if your block is successful, the momentum of the strike is still absorbed by your body and may knock you off balance.

Balance when moving

A fighter without footwork is like a wood worker with his tools nailed to the work bench. Without mobility, you will lose any advantage in both power and strategy. Mobility ties directly in with balance, power, and speed. You must therefore be careful with any motion that extends beyond the boundaries of your body. Overextending forward when throwing a punch, for example, may shift your center of gravity enough to unbalance you. Before we can learn about momentum, we must first learn how to move about in the fighting environment without losing balance.

Ease of movement:

1. Staying on the balls of your feet with your knees slightly bent is better than being flat footed, because it enables you to move faster and produce more momentum.

On balls of feet. **Flat footed.**

2. Ease of movement can be practiced in shadow boxing. Try to get a feel for techniques that allow you to keep a balanced state simultaneous to movement.

Linear movement:

1. Stepping with the foot closest to the direction of travel first eliminates the tendency to cross your feet. It also keeps you in your fighting stance and enables you to place your body weight behind your strikes.

2. Sometimes it is faster to step the way one would when walking or running. This can be done when you have a clear advantage over your opponent. But be aware that there is less stability in the middle of the step when your legs cross.

Circular movement:

1. When circling your opponent, try to remain in your stance as much as possible. A horse stance doesn't allow you the full benefit of your body weight, and a crossed stance makes you unstable when taking a blow.

2. If you are in a left stance and move your lead foot too much to the left, you will square your stance and be unable to place your body weight behind your strikes. If you move your lead foot too much to the right, you will end up in a crossed stance that is unstable, and that will inhibit ease of movement and prevent you from getting full reach in your techniques.

Switching stance:

1. When switching stance, you must move through the neutral stance, which also means that you will be less stable for a moment.

2. One of the best times to switch stance may be through other movement, as when throwing a kick with your rear leg and planting that foot forward.

3. Take advantage of your opponent by timing a counter-strike to land when he is in the middle of stepping or switching stance. This works especially well if he switches stance forward rather than back, because his forward step will add his momentum to yours.

Balance in throws and takedowns

A throw or takedown is a technique designed to get your opponent on the ground. The primary principle of a throw or takedown is balance. The most common types of throws are hip throws, where you rely on your hip to unbalance your opponent. But there are also neck throws, arm throws, body throws, and a variety of takedowns.

Balance in throws:

1. Throws that require a rearward shift of C.G. are the easiest to execute. This is because your opponent's body naturally doesn't bend backward very far.

2. Against a punch or grab attempt, apply a joint lock, simultaneously stepping behind your opponent's leg (the leg that is on the same side as the joint lock), and throwing him backward over your hip. Stepping behind the leg keeps him from shifting the C.G. for balance.

Rear hip throw with joint lock.

3. In a neck throw, place one hand on your opponent's chin and the other on the back of his head, simultaneously stepping behind his leg. Tilt his head back and to the side by pushing up on his jaw. Throw your opponent backward over your hip by pivoting your body.

Rear neck throw.

4. In a body throw, step behind your opponent with one foot, and place your arm across the front of his body. Throw him backward over your hip by pivoting your body, simultaneously applying downward focus. Your arm across his body helps to unbalance him by shifting his C.G. back.

5. In the hip throw, you only need to lift your opponent enough to get his feet off the ground. Once he starts going over your hip, his weight will begin to act to your advantage. His C.G. will be outside of his foundation and aiding the throw.

Balance in takedowns:

1. Takedowns can be executed through a joint lock, or by attacking your opponent's legs. In a leg takedown you will shift your opponent's foundation rather than his upper body C.G.

2. In a true self-defense situation, kicks to the front of the knee may be effective for unbalancing an opponent. In addition to stopping forward motion, the kick will straighten the leg and bring the opponent's upper body forward.

3. Leg sweeps, either toward or away from the centerline, can effectively unbalance an opponent. Sweeping toward his centerline may be the most effective, because this will also narrow his stance.

Balance when grappling

Once your opponent ends up on the ground, you can either finish the fight or walk away. However, one must also plan for a time when both fighters are on the ground. This is known as grappling. Grappling involves many positions, including prone, supine, on your side, on your knees, and straddling your opponent. You must study situations where you are in either the superior or inferior position. Just as in stand-up arts, your chances for success in grappling are greatest when you maintain your own balance, while simultaneously unbalancing your opponent.

Unbalancing an opponent on his knees:

1. Because joint locks work against the natural movement of the joint, these techniques can be used to unbalance your opponent in an attempt to reposition him. For example, if he is on his knees, a joint lock against the elbow can unbalance him forward to the prone position.

Center of Gravity   75

An opponent on his knees can be unbalanced forward with elbow leverage.

2. A technique that shifts the C.G. upward or backward can be used to take your opponent to the supine position. An example would be a palm against the chin to tilt the head back.

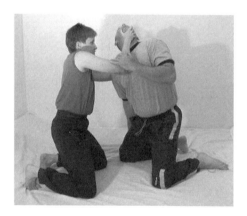

An opponent on his knees can be unbalanced backward with a palm to the chin.

3. An opponent on his knees can be taken to his stomach, back, or side, by employing circular motion in conjunction with a joint controlling technique. Because of your opponent's natural inertia, his body can't accelerate fast enough to keep his C.G. centered.

Unbalancing an opponent who is straddling you:

1. By raising your hips forcefully, you may succeed in shifting your opponent's C.G. upward and forward. If you simultaneously twist your body to the side, the move may have an even greater effect.

Raising your hips suddenly when your opponent is straddling you, may shift his center of gravity forward.

2. Any joint controlling technique used in conjunction with raising your hips will make it easier to unbalance your opponent.

3. Your opponent can also be unbalanced by shifting his C.G. down rather than up, but you must then employ circular motion simultaneously.

Pulling your opponent toward you, and simultaneously raising your hips and twisting his head, may unbalance him to the side.

Keeping yourself in a balanced state:

1. The principle of wide base and low C.G. applies. If on your knees only, be aware of that your base is narrower than if on all four.

2. If on all four and your weight is equally distributed over your hands and knees, just removing one point of support will be enough to disturb your balance.

3. If your opponent employs a joint controlling technique simultaneously, he can steal your balance even easier.

4. When straddling your opponent, stay in a low and wide stance. Be aware of that your opponent will take advantage of the moment you raise your C.G. to shift position or apply a different hold. Any move or shift in position must therefore happen explosively with an element of surprise.

Balance quiz

1. What are the advantages of the horse stance? Disadvantages?

Because of its wide base and low center of gravity, the horse stance is very stable. In addition, it allows you full focus on your hand techniques when practicing. But because the horse stance is somewhat rigid and wide, it limits your mobility. All targets on your centerline are also exposed to your opponent's strikes. The horse stance is therefore used more as a training stance than a fighting stance.

2. How can you increase stability?

You can always increase stability by widening your base and lowering your center of gravity. In a stand-up fight, bend your knees. In a ground fight, spread out.

3. When circling your opponent from a left fighting stance (left foot forward), which foot should you step with first and why?

When circling left, step with your lead foot first; when circling right, step with your rear foot first. This will keep you from narrowing your stance or crossing your feet, which is destructive to balance.

4. How do you keep from squaring your stance when circling an opponent? Why is it important to keep a sideways stance?

In order to maintain a sideways stance when circling, your rear foot must take a bigger step than your lead foot. This is because your rear foot is farther from the center of the circle. If circling right from a left fighting stance, and your steps are equal, you will eventually end up in a crossed stance. If the step with your rear foot is too big, however, you will eventually end up in a square stance. A square stance will expose your centerline and make you unable to keep the full weight of your body behind your strikes, which, in turn, will lessen power.

5. Why is the "girl push-up" nearly useless for building strength?

The farther the weight is from the pivotal point, the more strength is required to lift it. When doing push-ups on your knees, the pivotal point (your knees) is closer to the weight (your upper body) than when doing push-ups on your toes. As a result, you lift only half or less of your body weight. The regular push-up enables you to lift about 75 percent of your body weight.

6. Where is the center of gravity located on a horseshoe shaped object?

The center of gravity is always located toward the heavier end. On a horseshoe shaped object, the center of gravity is located in the open space toward the closed end of the arc. The center of gravity does not have to fall within the tangible parts of the object, only within the *center of mass*.

7. Why is a fighter throwing a side kick able to maintain balance, even though his base is very narrow and his center of gravity quite high?

Because the fighter's upper body is angled to the rear, his center of gravity is directly above his foundation (supporting foot). He is therefore in a balanced state. Any time you kick, you must center your upper body above your foundation.

8. Why is a punch or kick thrown with your lead hand (foot) naturally less powerful than the same strike thrown with your rear hand (foot)? Are there any advantages to throwing strikes with your lead hand (foot)?

Lead strikes have less distance available to build momentum for power. The advantage of the lead strike is that it is often faster than the rear, and therefore not telegraphed as easily.

9. How do you remedy the lesser power of the lead strike?

The power of the lead strike can be increased by taking a step forward. This increases the momentum by placing the weight of your body behind the strike.

Glossary

Acceleration--Change in speed and/or direction. An object that is in motion and changes its direction (a car driving up the clover leaf on-ramp to the highway, for example), will accelerate even if there is no change in speed. You can feel that acceleration is taking place by the way your body lurches forward, back, or sideways.

Balance--When the center of gravity falls above the foundation, an object is in a balanced state and will not tip over easily.

Basic Movement Theory--Whenever moving, you should step with the foot closest to the direction of travel first. This keeps you from crossing your feet and becoming unstable. Stepping with the foot closest to the direction of travel first also allows for smoother and quicker movement.

Center Of Gravity (same as center of mass)--The point in an object where all its weight seems to be focused. An object of uniform shape and weight has the center of gravity in the middle. An object of non-uniform shape and weight has the center of gravity toward the heavier end. To remain stable, the center of gravity should be as low as possible and above the foundation.

Force Per Square Inch--The narrower the base of an object, the greater the force per square inch. A narrow base is less stable than a wide base, but it also has more penetrating capability.

Gravity--An attraction between objects due to mass.

Hard Styles--These styles rely on wide stances and linear moves, with full focus of power into a single strike.

Horse Stance--A training stance simulating a person riding on a horse. The horse stance is used mostly for practicing basic strikes, allowing you to focus on your hand techniques. It also strengthens your legs and helps you maintain balance. When engaged in actual sparring, a fighter should use a modified horse stance that is not as rigid, but in which his knees are still bent, and his feet about shoulder width and a half apart.

Inertia--Resistance to change in motion. An object at rest tends to stay at rest; an object in motion tends to stay in motion.

Mass--The quantity of matter in an object. When acted upon by gravity, we can use mass interchangeably with weight.

Momentum--The product of the mass of an object and its velocity. The heavier the object, and the faster it travels, the greater the momentum.

Rotational Speed--The number of rotations per unit of time (revolutions per minute). In a spinning technique, the revolutions per minute in the center is the same as the revolutions per minute near the tip. However, the tip will move with a faster linear speed, because it is farther from the axis of rotation. The rotational speed is the same regardless of how far you are from the center, but the linear speed is proportional to the distance from the axis.

Sensory Overload--By striking and kicking with explosive combinations and to as many targets as possible in the shortest amount of time, you can create "sensory overload" and send your opponent into a state of confusion and chaos.

Soft Styles--These styles rely on flexible stances and circular moves, with the power building through one or more set-up strikes.

Torque--The product of the lever arm and the force. The longer the lever arm, the less force is needed to produce a given amount of torque.

Vector--An arrow symbolizing the strength and direction of a force. The longer the arrow, the stronger the force. For maximum power, all vectors must point in the direction of your strike.

Momentum
Without Movement, Nothing Happens

When I awoke this morning, I felt as though I had been sleeping on rocks. I tried to remember what we had been working on the previous day. It wasn't the kicks I had taken . . . rather, it was my instructor, who weighs 200 pounds, and with whom I had grappled. Because of my much lesser weight of only 120 pounds, I had occupied the bottom position during our grappling endeavor, trying to keep this enormous mountain of mass from squashing me into the floor. When I finally tapped out because I couldn't breathe, my instructor told me that in order to defeat him, I must rely on speed and explosiveness and work into a better position, from where I can attack his weaker areas.

We started the grappling exercise from a stand-up position. In order to grapple, you must first close in on your opponent. If distance is never allowed to lessen, the fight will never go to the ground. For a moment, my thoughts strayed to my first kick-boxing match. I had been matched against an opponent who was nearly equal in experience and weight. Two weeks before the match, I was notified that she had broken a couple of toes and couldn't fight. I had been training hard and was looking forward to testing my skill against the unknown. I asked the promoter if he could find me another fighter. He called me back three days before the match and asked if I would fight a lady out of my weight class. Her record was 5-0, and she weighed 155 pounds. At that time, I did not yet understand the importance of weight. In addition, the ring was small (16 X 16 feet), and there seemed to be only a fraction of a second before I found myself squashed up against the ropes, unable to press forward. If I had known then what I know now, I would have trained for position and not allowed her to get near enough to use her weight against me.

We all know that, if in the fight game long enough, we *will* get hit sometime during our martial arts career. Knowing this in advance, which of these two fighters would you rather take a punch from?

Most of us would choose the small skinny fighter over the big fat one, because he is the least massive of the two. But are you sure that a punch from the smaller fighter would hurt less? There is an interesting twist to this, which we will talk more about in Chapter 8 on Kinetic Energy.

Most people would agree that the mountain range separating Colorado and Utah is *massive*. Some people use *mass* interchangeably with weight. When we say that something is massive, we generally mean that it is heavy. In physics, the definition of mass is the *quantity of matter in an object*. If you take a massive object into space, it will become weightless because it is not acted upon by gravity. It will not become mass-less, however. Mass is also a measure of the *inertia* of the object. Inertia means *resistance to change in motion*. The greater the mass, the greater the inertia. This is important to power in the martial arts, because a fighter who possesses a large quantity of mass will be more difficult to stop than a fighter who possesses less mass.

Body mass in motion

If you can set a massive object into motion, it will take a lot of force to stop it. One of the principles behind powerful punching and kicking is therefore to keep the mass of your body behind the strikes. If you watch boxing, you will sometimes hear the commentator say that "he is an *arm-puncher*," meaning that the punches are thrown without the weight of the body behind them and will not do much damage. Your body has much more mass than your fist alone, and if the strike can originate in the motion of your body, you can generate a great deal of power. This can be accomplished by taking a step forward, or by pivoting your foot, hip, and upper body in the direction of the strike. An added benefit of stepping or pivoting is that it increases your reach. You can therefore land a strike that initially appears to be too far from the target. The drawback, of course, is that stepping takes time. In the pictures below, Keith Livingston (former P.K.A. Utah champion, and I.K.A. light heavyweight Rocky Mountain champion) demonstrates how pivoting your foot, hip, and upper body increases your reach.

Caution: Be aware of your opponent's timing, so that he doesn't counter your forward step with a strike. Your forward motion would then increase the power of your *opponent's* punch, and would therefore not work to your benefit. This is often referred to as ***adding momentums***.

But what if you want to throw a strike that requires a sideways or slightly circular motion, like the hook or round house kick? How do you place the mass of your body behind such a blow? For best effect, the mass of your body must move in the exact same direction as the strike.

Note how Keith pivots on the ball of his foot to place the mass of his body behind the hook.

Note: If you hear your instructor say "dig for power," he means for you to place more weight on the foot that is on the same side as the punching arm. This is not the same as being flat-footed. A flat-footed fighter is not as mobile as one who is up on the balls of his feet. To dig simply means to *place more weight per square inch* on the foot.

The power of a round house kick can be increased by taking a lateral step with your supporting foot. You must also allow your upper body to pivot in the same direction as the kick. If your upper body is rigid, you will not be able to bring your hips through, and the kick will almost follow the vertical path of a front kick.

In order to place your body mass behind a strike that follows a vertical path upward, like the uppercut, you must keep your legs slightly bent prior to striking. This allows you to spring from your knees, placing the weight of your body behind the blow.

Note how Keith keeps his legs slightly bent when throwing the uppercut, and how he pushes off against the floor with the ball of his rear foot for power.

The last of the four major directions in which a strike can be thrown is downward. Your body weight must now drop for power. An added benefit is that the force of *gravity* will work to your advantage.

Martina drops her weight for momentum when throwing the downward elbow strike.

So, you see, weight is beneficial because of the *momentum* it can produce. If you can utilize the weight of your body, your strikes will be more powerful than if you rely on muscle strength alone. Momentum is also increased whenever you add motion in the direction of the strike.

Relying on the entire weight of your body increases the power, but it also takes more energy than throwing with the arm only, so there is a trade-off. Also, be careful not to jam your own strike by getting too close to your opponent before unleashing it.

In order to add body momentum, the technique must start in your lower body. When we get tired, we often feel as though our legs are lagging behind. This is also a reason why leg strength is so important: The legs are used to advance you forward. If the motion starts in your upper body, you will be unable to bring power to the strike, because the upper body will be working on "dragging the lower body along."

We have now looked at how to keep the mass of your body behind a strike that is thrown forward, sideways, upward, and downward. There are, of course, several variations with angles that fall in between these four major ones, including spinning techniques which rely on both linear and circular motion. This is where we need to employ the concept of *vectors* and *resultants*, and the concept of *rotational inertia*, which we will discuss in detail in Chapters 4 and 5.

Working with momentum

Shortly after I started in kick-boxing, my instructor asked me to hold a kicking shield so that he could demonstrate the stepping side kick to the rest of the class. I grabbed the shield and watched him take a few steps back for distance. The next thing I remember is being sprawled out on my back on the floor, my body still tingling from the power of his kick. "If you were going to run through a closed door," he said, "you wouldn't start running, then come to a stop in front of the door, and then bump your shoulder into it, would you?"

In order too increase the power in your strikes, you must add motion. But any time you start an object in motion and then attempt to stop it, you must overcome inertia. When you restart the object in motion, you must again overcome inertia. Overcoming inertia takes energy and is destructive to power. So once you have started a strike in motion, you must continue the motion through the target with no stopping in between.

> *Momentum is defined as the product of the mass of an object and its velocity.*
> **Momentum = mv**

Velocity is often used interchangeably with speed. The difference is that velocity also has *direction*, while speed is simply a measure of how fast something is going. Take a look at the following equation:

Momentum = mass (weight) X velocity (speed). Let's say, for simplicity, that we give the momentum the number 1000. Let's also say that we give the fighter a weight of 200. What would his speed be? When you plug these numbers into the equation, you will see that you must time the weight by 5 in order to arrive at 1000. We can therefore say that: **1000 = 200 X 5**. If the fighter weighed less, say 125 instead, his speed would have to increase in order to reach the same momentum: **1000 = 125 X 8**. So, a lightweight fighter must move faster in order to achieve the same momentum as a heavyweight.

If you are asked to run a distance of five yards as fast as you can, and then run a distance of one-hundred yards as fast as you can, you are likely to reach your highest speed somewhere in the second run. This is because you have more time and distance in which to accelerate. This greater speed will give you more momentum, which will make you more difficult to stop.

The power of a strike will increase if it has a long distance over which to build speed. Test this concept by throwing a punch with your arm already half extended. Next, throw a punch from your shoulder. The additional distance to your target will give your fist more time to build momentum. This is also why kicks thrown with the rear leg usually seem more powerful than kicks thrown with the lead leg: the distance to the target is longer.

Are there any exceptions to this? Of course there are. Take the uppercut, for example. Assuming that the strike is thrown mechanically correct, which uppercut will be more powerful, the lead or the rear? According to the principle above, the rear uppercut should be more powerful. However, with the uppercut, the *distance of importance* between your fist and the target is not horizontal distance but vertical distance. Because your rear hand is farther horizontally from the target than your lead hand, you will not be able to place as much body mass behind the punch *vertically* as you will with the lead uppercut. A rear uppercut thrown vertically straight will miss its target, because it will not be lined up with your opponent's chin. A rear uppercut must therefore be thrown at a slightly diagonal angle upward. The lead uppercut, however, is lined up (or nearly lined up) with your opponent's chin, and allows you to place the full weight of your body behind it.

Because of its vertical path, the lead uppercut allows you to keep the full weight of your body behind the strike. →

← **The rear uppercut is farther from the target and must be thrown at a slightly diagonal angle upward.**

Lowering your weight improves your balance (low center of gravity and wide base = stability). With a strike that is angled upward, like the uppercut, lowering your weight prior to throwing the strike also allows you to push off against the floor for explosiveness.

Cindy Varnell Winlaw drops slightly in the knees prior to throwing the uppercut. This helps her use the weight of her body to increase the power of the strike.

Lowering your weight is great for combining defense with offense, creating more power through your defensive move. Let's say that your opponent throws a punch, which you duck by bending at the knees. Naturally, you have to come back up again, but you now take advantage of your defensive move and throw an uppercut *within* the upward movement your body has to make in order to reset into your fighting stance. This is what I refer to as "catching two flies with one swat"; defense creates offense.

Martina ducks her opponent's strike and uses the necessity to reset in her stance to launch an uppercut.

Momentum   89

The principles of power often seem contradictory. For example, it is better to be heavy than to be light, but it is also better to be fast than to be slow. A heavyweight has more difficulty gaining speed than a lightweight. If you are heavier than your opponent, you should try to use that extra weight to your advantage without letting it become a burden.

The pictures below are a demonstration of how you can benefit from your weight rather than from your speed. Keith Livingston is relying on the forward momentum of his body (delivering the kick with his shin, with full body weight behind it), instead of the speed and snap of his leg. Note also the jump, which makes it easier to advance forward with explosiveness.

**Momentum = mass X velocity
or
Momentum = mv**

Momentum = $\mathbf{m}$v

Keith delivers a round house kick with the shin.

A heavyweight usually has the advantage at close range, where the lightweight has difficulties out-muscling him. A strike that is thrown from too tight a distance, however, will lack extension, and power will be stifled. A strike that is thrown from too far out will lack penetrating force. When learning to judge proper distance for your hand techniques, pay attention to the length of your arms, and the fact that a strike that is *mechanically correct* with the weight of the body behind it, will have better reach than one which is not. Tiny adjustments in distance will be necessary throughout the constantly dynamic fight.

A different look at distance and reach

We normally think of distance as the actual footage between ourselves and our opponent, or between our striking weapon and the target. We also tend to equate distance to safety. We know that if we are within reach of the opponent's arms or legs, we are also in danger of getting hit. Likewise, if the opponent is within reach of our arms or legs, we have a chance of landing a strike. Thus, distance can be decreased or increased by taking a step forward or back.

Staying outside of the opponent's striking zone is beneficial when you are tired or need time to calculate your next move. But there are also disadvantages to having a lot of distance. Moving from long range to close range is time and energy consuming, and using a lot of motion from far out may telegraph your techniques. You must also consider your opponent's distance. If he is out of reach, should you "chase" him, or is it better to wait until he makes the first move? There is, of course, no definite answer, as much of it depends on your strategy. If you chase him, you may be able to land a strike while he is backpedaling and can't effect a powerful counter-strike. But if you wait for him to make the first move, you may have a better chance of increasing the power in your strike through timing. In general, it can be said that distance should be closed when your opponent experiences a moment of weakness, like right after you have successfully defended and countered one of his long range techniques. Distance can also be used strategically to lure your opponent forward and into your strike.

The most lucrative fighting distance may be where you are far enough that your opponent can't reach you, yet close enough that you can reach him. This sounds like a contradiction. Yet, it is possible to manipulate distance, and therefore reach, without changing the actual footage to your opponent.

Finding your optimal reach: The optimal reach for a punch is horizontally straight out from your shoulder, and the optimal reach for a kick is horizontally straight out from your hip. Any time you strike or kick above or below the horizontal, your reach will decrease. In practical terms, this means that if your target is high or low, you must be closer to your opponent in order to land a strike. Try the following:

1. Face your partner and, without moving, extend your arm horizontally toward him. Adjust your distance forward or back until your fist barely touches him. With all other factors constant, this is your optimal reach.

2. Without moving, raise your arm toward your partner's face. Note how you can no longer reach him. Now lower your arm toward his midsection. Again, note how you can no longer reach him. Even though your position and the length of your arm has not changed, your reach has.

Does this mean that you must sacrifice safety for reach whenever striking above or below the horizontal? Not necessarily, but you may want to use your hands when striking to high targets, and your legs when striking to low targets, as this will keep your arm or leg as close to the horizontal plane as possible, allowing you to work from a greater distance.

A strike that is extended horizontally from your shoulder has the greatest reach. Striking above or below the horizontal decreases your reach by several inches.

Distance may also vary depending on which target you strike. For example, the distance between your fist and your opponent's chest is longer than the distance between your lead foot and your opponent's lead knee. You may therefore want to choose targets that are the most appropriate for the strike (or strikes that are the most appropriate for the target). Now, take this concept a step further and apply it to defense. Try blocking whatever comes below the waist with your legs, and whatever comes above the waist with your arms. This gives you minimal movement, allowing you to conserve energy and speed up your counter-strikes.

92 Fighting Science

How to increase your reach: As you can see, reach is very important. Unfortunately, we can't do much about the length of our arms and legs. The good thing is that there are ways to increase reach without stepping closer or growing longer limbs.

Crouching increases your reach. This can be tested on the heavy bag. From an upright fighting stance, extend your arm straight toward the bag, with your knuckles barely touching. Without stepping, crouch by bending at the knees. Did your reach increase by two to three inches? This happened because your upper body automatically moved forward when you bent at the knees, placing your shoulders closer to the target. Note that crouching does not imply bending at the waist with straight legs, as this may impair your balance or expose your facial area to kicks.

Crouching increases your reach significantly. When your upper body moves forward, momentum is also increased.

Pivoting increases your reach. Pivoting, without pushing off or leaning toward the target, increases your reach by three to four inches by extending your shoulder (or hip, if kicking) toward the target. An additional benefit is that pivoting places the weight of your body behind the strike, and therefore increases the power.

Pivoting increases your reach significantly.

Pushing off increases your reach. Elongating your body toward the target by pushing off with your rear foot can increase your reach by one to two feet. Reach can now be used strategically by circling your opponent at a distance outside of the apparent danger zone. This allows you to stay outside of his striking zone and still land your strikes. When throwing the punch, be careful with balance, so that you don't overextend your center of gravity.

By pushing off with her rear foot and elongating her body toward the target, Martina is able to increase her reach by a foot or more.

94   Fighting Science

Once you understand how to increase your reach, you can now set your opponent up to believe that you are too far away to strike him.

Using distance deceptively: The most interesting part about distance is perhaps how you can use it coupled with timing to deceive your opponent into thinking that he is safe. There are two types of timing:

1. The timing of your strike to your opponent's movement and distance.

2. The timing of your movement and distance to your opponent's counter-strike.

Fighting, at this stage, becomes very complex, where a distance of three feet is not a distance of three feet at all times. The concept of *half-stepping* can give your opponent the perception that you are increasing the distance between you. First, take a step to the rear with your rear foot only. This will widen your stance, move your upper body slightly to the rear, and create the illusion that you have increased distance. Your opponent now feels it is safe to come forward. When he does, you can simultaneously step forward with your rear foot and strike.

If the concept of half-stepping is applied to the elongated lead strike, the strike can be landed from an excess distance of two to three feet. First, take a big step forward with your lead foot only. This will widen your stance. Simultaneously push off with your rear foot and stretch your body by pivoting your hip and shoulder in the direction of the strike. Your reach has now increased significantly. After the strike has landed, bring your lead foot back to reset your body's balance and get back to long range.

Taking a step forward with the lead foot, and elongating the body toward the target, enables a smaller fighter to fight outside of her opponent's reach and still land strikes.

If you get too close to your opponent, you might smother your own techniques to where you can't strike effectively. You can now increase distance without stepping by centering your weight above your rear foot. This moves your upper body slightly to the rear, but without moving your foundation. After you have thrown the strike, move your upper body forward again to smother your opponent's counter-strikes.

Note: Any move that places your weight above your rear foot should be brief and timed to a specific strike. You are naturally more vulnerable when your weight is to the rear.

So, you see, having full control of distance allows you to move out of reach of an aggressive opponent, or to move to close range and smother his long range techniques. You can also move to a superior position to his side and strike when he least expects it. Once you understand how to manipulate distance to maximize your reach, you may want to study how movement is detected, so that you can defend and counter your opponent's attack at the earliest stage without wasting energy.

Reading between the lines: One of the difficulties with judging distance is that we tend to judge to the "wrong" target. We see what is closest to us. For example, instead of measuring the distance to our opponent's jaw or nose (which is usually a foot or more behind his guard), we will measure the distance to our opponent's hands. That is because his hands are the first barrier; it is what is closest to us, and what we are the most concerned with. This gives us a tendency to strike our opponent's hands, even if we have enough reach to strike his face. Think of it as reading a book. You will see the black print on the white page, because this is what "stands out" much like the barrier of your opponent's hands. When training for target accuracy and power, you must train to "read between the lines"; to see what *is not* there.

Speeding up your punches

When the instructor paired us up to spar in karate class, he told my opponent: "Watch out for her; she's got mongoose blood in her!" Later, when I started in kick-boxing, my instructor sneered and said that he saw my kicks coming three days ago, and that my punches were so slow it's like "waiting for water to boil."

The speed of your hand techniques depends on many factors. Some people's physical makeup and inherited abilities enable them to throw faster punches than others. But, like most things, speeding up your techniques is mostly a learned trait, which can be attained with diligent practice.

Some factors that need to be considered when working on your hand speed are:

- **Economy of motion.** The less wasted motion, the faster the strike. Avoid throwing the strike wide, or pulling back prior to throwing the strike. The "pulley effect" is also part of the economy of motion concept. As one hand starts on its way back, the other hand should start on its way out. This will have the effect of your hands helping one another and will decrease the beat between strikes.

The pulley effect can be better understood with the aid of a rope. Wrap the rope behind your neck or body and grab one end with each hand. When your left hand is extended in a strike, your right hand is all the way back at the guard position. As your left hand starts to retract, your right hand starts its forward motion. They meet at the halfway mark. This is the most economical way of striking, with the shortest beat between strikes, without sacrificing body mechanics.

- **Overcoming inertia.** It takes energy to start your fist in motion, but it also takes energy to stop it once it is moving. The faster you can throw your *initial* strike, the faster the rest of the strikes in your combination will be. This is because the first strike tends to set the pace. It takes more energy to increase the speed throughout a combination than to simply maintain a speed that has already been set.

- **Avoid falling into opponent's rhythm.** In regards to power, the primary reason for wanting to increase the speed in your strikes is because a faster strike has more momentum. A secondary reason is because it allows you to overwhelm your opponent and beat him to the opening. It is easy to fall into your opponent's rhythm, however. If he throws his strikes at a certain speed, you must defend against them by blocking or bobbing and weaving at that same speed. But as soon as you have finished your defense, you must speed up your offense. You will now be working with two speeds: a slower speed that is in tune with your opponent's speed and enables you to defend against his strikes, and a faster speed for your offense, which enables you to beat your opponent to the opening.

One thing that often seems to inhibit hand speed is the inability to relax in the upper back and shoulders. Whenever we get ready to strike, the body has a natural tendency to tense the muscles involved in throwing the punch. This tensing stifles speed and keeps us from sending the power of the punch through the target. In other words, we will hit the target but stop short power-wise. The punch will feel more like a push.

Note: Because of the *impulse*, a push is not as powerful as a punch that is snappy. Impulse is defined as the *change in momentum*. We will learn about this in Chapter 6.

The reversal of motion, as when pulling the hand back in defense or in preparation for a second blow, is also more difficult when the muscles in the back and shoulders are tense. Because it takes energy to stop a punch in motion, and because a punch must be stopped completely before it can reverse direction, it becomes especially important to stay relaxed. A punch that is thrown truly relaxed is thrown with ease and with the punching arm almost fully extended at impact, and will seem effortless.

Caution: Staying relaxed in the shoulders does not mean to allow your hands to drop. Your guard must stay high for protection. When shadow boxing, pay attention to where your shoulders and back feel strained or tense. Work on keeping your shoulders down and relaxed, yet keeping your guard up.

When your opponent is unable to defend against the punches thrown at him, *sensory overload* will occur. This usually happens when throwing lengthy combinations that are consistently faster than your opponent's. Staying relaxed will help you throw your combinations in spurts to create this effect. Blocking all strikes at this level is almost impossible, and some of your strikes will land.

Another interesting concept is that there is usually one fighter dominating the fight through speed, and the fighter who is slower will subconsciously resign to the fact and be unable to bring his strikes up to par. If this is brought to the slower fighter's attention, either through a trainer or through his own mental determination, and he suddenly increases the speed in his strikes, the fighter who was originally faster will now start getting hit. As a result of getting hit, he will have a tendency to slow his strikes to the pace his opponent was at earlier during the fight (this is more of a mental issue

98   Fighting Science

than a physical one). This may well reverse the domination of the fight from the faster fighter to the slower fighter (who has now become the faster fighter). If you are originally faster than your opponent, and your opponent suddenly picks up speed, you must increase your own speed to keep him from taking the fight from you.

The power of the sideways stance

When moving or stepping forward to close distance, we sometimes have a tendency to square our stance. This happens because we unknowingly fidget or take many tiny steps when stepping. When it is time to strike, the punches lack power because the weight of the body is not behind them. The fist is connected to the arm, which is connected to the shoulder and, in a horse stance, only a very small portion of the body weight is directly behind the punch (basically just the weight of the shoulder), whereas in a sideways stance, most of the body weight is behind the punch.

The fighter on the left does not have the full weight of his body behind the blow, and can therefore not generate as much power as the fighter on the right.

Which of the following three fighters will be able to generate the most powerful punch? Why?

The fighter on the left carries his elbows behind his body and can therefore not utilize the full benefit of his body mass. In addition, he is leaving his head open for blows. The fighter in the middle carries his elbows too far from his center of mass and is leaving the sides of his body exposed. The fighter on the right is in a good sideways stance with his guard high for protection, and his elbows in front of his body for power.

Adding momentums

Now, when you have learned how to use your mass and velocity against your opponent, you should learn to use your *opponent's* momentum against him. Which would have more impact: driving a car at 30 miles per hour into a stationary object, or driving a car at 30 miles per hour into another car that is moving toward you, also at 30 miles per hour? Which would have more impact: getting hit by your opponent's straight right, or walking forward and into your opponent's straight right?

Keith pulls his opponent forward and into the knee strike, relying on both his own and his opponent's momentum.

In the picture below, Cindy slips her opponent's punch, simultaneously moving in with a jab to the midsection. This is another example of adding momentums, or what I call "catching two flies with one swat."

In the picture above, Cindy uses the principles of physics to accomplish defense and offense at the same time. She has also timed the strike to connect when her opponent is on one leg, either in the process of stepping or kicking. Being on one leg means that your base is narrow and your stance is not as stable. This is a very bad time to absorb the power of a strike.

Summary and review

> **Momentum = mass X velocity**

In terms of physics, momentum is the mass times the velocity. We can also think of this as the weight of your body times how fast you are moving. The heavier you are, or the faster you move, the greater the momentum. The principles of momentum state that:

- There must be some motion in order to achieve momentum.
- To benefit from the motion, it must be in the direction of the strike.
- The more weight you can place behind the strike, the greater the momentum.
- Because the purpose of momentum in the martial arts is to move your opponent back or to knock him over, your weight is more important than your speed.

Momentum when striking

Momentum has the ability to move your opponent through a distance, or to knock him down. Because mass is a significant element of the momentum equation, it is important to place the weight of your body behind all strikes. Let's talk about weight first, and look at the element of speed later.

Straight strikes:

1. A strike is considered straight if it shoots straight forward or back along the centerline of your body. Examples of straight strikes are jab, rear cross, reverse punch, heel palm strike, and back elbow strike. Variations of straight strikes are reverse downward hammer fist, and forearm strike or ridge hand from an offset position to your opponent's side.

2. The mass of the body must be behind the strike. This is accomplished by pivoting the hips and body. The movement must start in the body, or the strike will work on "pulling the heavier body along."

A strike without body rotation uses arm power only (left). A strike with body rotation has more momentum derived from the weight of the body (right).

Momentum can also be increased by pulling your opponent into the strike, as in the elbow sandwich (left), or by pulling on his arm and simultaneously throwing a palm strike to his jaw (right).

3. For good balance, both feet should be planted and push off against the floor. Leaning in the direction of a strike adds the weight of your body to the blow, but you must be careful with balance. You don't want to lean so that you are on one foot only.

4. You can multiply your effective power many times by relying on the weight of your body rather than just the arm. At the initiation of a strike, your elbow should be held in front of your body. This allows you to use your heavier and stronger body to start the elbow and fist in motion.

If your elbow is behind your body at the initiation of the strike (left), it must start its motion first, and will spend energy on dragging the heavier body along.

5. Momentum can be increased through the "push-pull" principle. It is not just the striking side of your body that brings momentum. "Pulling" with the opposite side simultaneous to "pushing" with the striking side allows your body maximum follow-through.

Striking without using the push-pull principle doesn't allow you as much penetration (left), as striking with the push-pull principle (right).

6. Additional momentum can be gained by taking a step in the direction of the strike. Whenever possible, step forward rather than back. If you have to step back, then try to time your strike to land as your opponent steps toward you. This allows you to take advantage of his momentum.

7. When throwing double strikes with the same hand, it is easy to cheat and reset halfway only. But, because you don't have the benefit of distance, this is detrimental to power. If you reset halfway only, then you should also take a step forward with your next strike to create momentum.

Looping strikes:

1. A strike is considered looping if it follows a slightly curved path and impacts from the side rather than straight to the centerline. Examples of looping strikes are hook, horizontal elbow, shuto from the side, ridge hand, hammer fist, soft palm strike or slap from the side.

2. The back fist loops away from your centerline rather than toward it. Because this goes against the natural movement of the joint, you must be especially careful to rely on the rotation in your body for momentum.

3. Power in the looping strike is attained by pivoting in the direction of the strike with all moves synchronized, or by taking a lateral step in the direction of strike.

4. Looping strikes that follow a short looping path (hook, elbow) should rely on body mass for power. This is because they don't have the benefit of building power through distance. A short step in the direction of the strike can also be used.

5. Be careful not to lean into the looping strike, as this would shorten the distance and starve the strike of power. This is especially important with short looping strikes, like hooks and elbows. Your balance may also suffer. It is better to set down (lower your weight) with the strike.

6. Regardless of whether you throw a straight or looping strike, impact must be straight for maximum force. If the strike is sliding against the target, maximum power cannot be attained. When you hook, don't pull the strike back toward you while impact is still in progress.

7. When throwing alternating looping strikes, use body momentum from the return of the first strike to launch the second strike. This is a variation of the push-pull principle.

8. Any looping strike can take advantage of your opponent's lateral movement. If your opponent steps toward your right, throw a right strike, and vice versa. He will now be walking into the strike. This also makes fighting economical, because you rely on your opponent's movement rather than your own.

9. Hammer fists, shutos, and ridge hands can also be thrown with a step forward for momentum. Because these strikes come from the side, stepping past your opponent allows the momentum of your body to continue past the target.

Ridge hand with body momentum continuing past the target.

Upward and downward strikes:

1. Most upward strikes follow a path along the fighter's centerline. Some upward strikes follow a slightly diagonal path. Examples of upward strikes are uppercut, upward elbow, upward palm strike, and x-strike.

2. A downward strike is thrown with the aid of gravity. Examples of downward strikes are downward elbow, downward shuto, downward hammer fist, overhand strike, and x-strike.

3. Any strike that follows an upward or downward path must have the weight of the body behind it. In the upward strike, your knees should be slightly bent, allowing you to push off against the floor. The opposite is true for the downward strike. You must now start in a higher position and drop your weight.

4. When throwing the upward strike, pay attention to balance. Because the strike follows an upward path, it is easy to raise your body up and become top heavy.

Momentum when kicking

Like striking, momentum when kicking is gained by placing the weight of your body behind the kick. There are numerous ways this can be done. The important part to remember is that both your weight and movement must be in the direction of the kick. Physics and martial arts is not just about how to apply the principles of physics, but also *when* to apply them. There are times when one principle works better than another, or when you can gain an advantage by allowing your opponent to use the principles of physics.

Momentum in different direction kicks:

1. Momentum in straight kicks (front kick, side kick, front knee) is gained by pivoting your hips, or by taking a step forward. The step and kick should be synchronized into one fluid motion. Breaking the technique into two moves halts the momentum.

2. Momentum in looping kicks (round house, crescent, round knee, side knee, hook kick) is gained by pivoting your hips, or by taking a step to the side and in the direction of the kick. When throwing a kick with a path away from your centerline (outside crescent kick), you must allow your hip to lead, or the movement will "exhaust" itself.

In the outside crescent kick, the heavier body should lead. Note how the leg remains chambered until the body has achieved rotation.

3. Momentum in downward kicks (axe kick, stomp) is gained by dropping your weight with the kick, relying both on your body weight and on gravity.

4. Momentum in jump kicks is gained through movement in the direction of the kick, and by contracting your body for faster rotation.

The importance of bending the knees:

1. Keeping the supporting knee slightly bent when kicking enables you to push off against the floor and spring from your knees. This will place your body weight behind the kick for momentum.

2. The kicking leg should also be slightly bent until impact is complete. If your leg is absolutely straight on impact, you won't have anything left for penetrating the target. Your leg should be positioned so that you can hit and extend through the target. This also requires correct judgement of distance.

Lead leg versus rear leg:

1. The lead leg is faster, but the rear leg has a longer distance for building momentum. If the lead leg is used, the hips may be pushed forward (as in a front push kick).

Front push kick off lead leg.
Momentum is increased by pushing the hips forward.

2. Whenever you throw a kick with the lead leg from a stationary position, you will be slightly out of reach. This is because your upper body must move above your rear leg for balance. If you take a step forward with your rear foot prior to throwing

the lead kick, you will keep your balance, increase the momentum, and attain better reach.

3. Momentum in the spinning back kick can be gained by allowing your body to move forward above your supporting leg on impact. Your reach will increase by half the distance between your feet prior to kicking.

The weight of your body can be added to a spinning back kick by allowing your upper body to move forward and above your supporting foot.

Increasing momentum:

1. Momentum can be increased by taking a step in the direction of the kick. Stepping with the non-kicking leg is quicker than stepping with the kicking leg.

2. When throwing a lead leg kick, stepping with your rear leg first will center your foundation under your upper body for balance.

3. When alternating lead and rear kicks, you will automatically advance forward in a manner similar to walking. This gives you the smoothest movement, but not necessarily the best protection.

4. The power of the round house kick can be increased by taking a lateral step in the direction of the kick.

5. When kicking with the intent to knock your opponent back, use a kick that follows a straight path: side kick, front kick, etc. The same is true when adding

momentums. For greatest effect, your strike or kick should be in the exact opposite direction of your opponent's movement.

6. If you throw two front kicks in a row with your lead leg, you must take an adjustment step forward with your rear foot between kicks. This is because a straight kick is likely to set your opponent back.

7. The side kick can be thrown with either your lead or rear leg. When kicking with your rear leg, you can gain momentum without a step. This is because your lead foot becomes the pivot point, with your body moving forward at the initiation of the kick.

A side kick with the rear leg does not require a step for momentum.

8. In a sliding kick, additional momentum is gained because your body is still moving forward on impact.

9. When kneeing your opponent (without grabbing), momentum and power is gained through a fast gap closure.

Force per square inch:

1. The smaller the impact weapon, the greater the force per square inch. An example is the axe kick, impacting with the heel. Work on bringing your leg as high as possible, so that you can use gravity and distance to your advantage. Also use the muscles in your leg, together with your body weight, to pull the kick down and through the target.

110   Fighting Science

The axe kick has a great force per square inch, and can be dropped on the nerves of your opponent's thigh.

2. The knee strike has a lot of power, because of the small surface area of the knee. Because the knee strike is thrown from close range, you will have to rely on body weight without the benefit of momentum through distance. Grabbing and pulling your opponent into the strike increases the momentum.

Jumping and kicking:

1. Momentum can be increased by jumping simultaneous to kicking. A jump allows you to move both feet freely through the air without the problem of friction.

2. When jumping and kicking, your will be able to spin faster if you vary the shape of your body. To avoid splitting the resultant force, a jump kick should occur at the apex of the jump.

3. Jump front kicks, round house kicks, and side kicks, too, should rely on a compact body. If one leg is pointed toward the ground, and the other toward the target, you can't direct your energy properly into the target, and will split the resultant force.

Momentum through natural combinations:

1. Momentum can be gained by allowing one kick to build speed for the next. A round house kick can be thrown prior to a spinning back kick, because they both follow the same path, with the second kick flowing naturally from the first. The momentum of the first kick can be used to accelerate the second kick.

2. When throwing multiple knee strikes, momentum is gained by kneeing with your rear knee first, setting that foot down forward, and throwing your next strike with your other (rear) knee. This will also produce momentum that will move your opponent back.

Multiple alternating knee strike, setting the kneeing leg down forward, will increase the momentum through distance.

The resultant force:

1. When throwing the outside crescent kick, your body should lead, or the movement will exhaust itself, and you will be relying on leg power alone.

2. A common tendency with the outside crescent kick is to kick upward upon impact. But this would split the power with some going into the target and some toward the ceiling. The power of the outside crescent kick should go horizontally through the target.

Kicking on the ground:

When kicking on the ground, the pivot point will shift from the foot to the knees, hands, sides, or elbows.

1. Almost every kick that can be thrown when standing can be thrown when on the ground. The only change is the pivot point. Instead of pivoting on your supporting foot, you will pivot on your supporting knee. Kicks may also be thrown when supported on your back, side, or elbows.

2. When kicking on the ground, momentum is achieved by rolling onto your side (as when throwing a round house kick), or by raising up on your elbows (as when dropping down with an axe kick).

3. When on the ground, movement can be used to create momentum for getting back on your feet. A back kick or side kick can be timed to your opponent's advance. When the kick sets him back, do a forward roll to get to your feet.

A back kick from the ground can stop your opponent's advance. Remember that your greatest reach is horizontally from your hip. If you kick at an upward angle, your opponent must be closer in order for the kick to land (left). This requires correct timing. After impact, a forward roll will give you momentum to gain distance and get back to your feet (right).

Momentum in defense

In defense, momentum can be used to launch a counter-attack, to defend against multiple strikes in a very short time, or to avoid a strike through movement.

Principles of linear blocks:

1. Linear blocks meet the strike straight on. Examples of linear blocks are upward block, downward block, and shin block. Moving your body in the direction of the block allows you to use momentum to your advantage, and may knock your opponent off balance.

2. Meeting the strike with your block makes it more difficult for the strike to penetrate your defense. Your forward momentum will cancel some of your opponent's momentum.

Momentum   113

Principles of redirecting blocks:

1. Redirecting blocks allow your opponent's strike to continue past the target. Examples of redirecting blocks are inward and outward blocks and parries.

2. Don't reach for the strike with your block; keep your arm tight and use your body weight. Moving your body laterally toward the strike will have the effect of knocking the strike off the path of power.

The redirecting block connects at an angle perpendicular to the strike, allowing the strike to continue past the target.

Principles of evasive movement:

1. Slipping or bobbing and weaving can be used to gain speed, and therefore momentum, for a counter-strike.

A bob and weave can be used to gain momentum for a hook. Use strikes that naturally follow the same path as the evasive movement, so that there is no stop in momentum.

2. Pivoting off the attack line allows the momentum of your opponent's strike to continue. Try to time a counter-strike or kick to his forward motion.

114   Fighting Science

Take advantage of your opponent's momentum by allowing it to continue past you and into your counter-strike. A round house kick to the midsection may be a good technique.

Body weight behind blocks:

1. A forceful block relies on the same principle as a forceful strike: use your body weight in the same direction as the block.

2. When blocking against an overhead strike, meet the strike slightly by stepping forward, but without raising up in your stance.

3. When blocking a kick, meet the kick slightly by dropping your weight with the block. The elbow is a vicious striking weapon, but many fighters forget that it is an equally vicious blocking weapon. The elbow is very sharp and strong and has the ability to focus the power over a very small surface area, inflicting great pain. Dropping your weight increases the effectiveness of the elbow block.

4. Stepping toward your opponent's strike simultaneous to blocking increases the momentum and may knock your opponent off balance.

Momentum and defense create offense:

1. A defensive move can be used to build momentum for a counter-attack. For example, slip your opponent's strike with a rearward slip. Counter-strike when your head and upper body come forward. This increases the momentum through forward motion and body weight.

Momentum 115

Rearward slip and counter-strike.

2. When blocking or parrying and counter-striking, it is more effective to counter-strike with the non-parrying hand than with the parrying hand. There are two reasons for this: first, your body will be chambered for a counter-strike with the opposite hand. Second, this allows you to decrease the beat between parry and counter-strike.

Parrying and countering with the same hand.

Parrying and countering with the opposite hand.

Momentum through circular motion:

1. The longer the distance to the target, the more time you have to build momentum. Circular motion allows you to gain speed, and therefore power. Circular motion also allows you to block simultaneously with both hands without contradicting body mechanics. An example would be a downward parry with one hand and an outward block with the other hand.

2. Evasive movement can also rely on circles. An example is bobbing and weaving. Take advantage of this by using the circle to accelerate a counter-attack.

Momentum in throws and takedowns

A throw is generally more violent than a takedown, and therefore has a potentially higher success rate. But because of a variety of factors, including resistance from the opposing party, difficulty to attain correct positioning, etc., throws don't always work. Depending on how the situation unfolds, and as long as you understand that both throws and takedowns utilize the same principles of motion, you can use either, with the main difference being positioning. A throw requires that you are very close to your opponent (touching), while a takedown may permit a little more space between you. In addition, a throw generally utilizes a tight circular motion over your hip or body, while a takedown utilizes a larger circle around your body.

Momentum and balance:

1. In a combative situation, the adrenaline rush may make you forget the small intermediate steps that matter most. It is therefore important to steal your opponent's balance before attempting the takedown.

2. If you start the takedown or throw by stealing balance, your opponent will be unable to counter until he hits the ground.

Push-pull principle in throws:

1. Twisting your body, using the push-pull principle, will make a takedown or throw more dynamic. For example, get hip to hip with your opponent (facing opposite directions), and place one hand on each of his shoulders. Push against

his shoulder or chest with one hand or forearm, simultaneously pulling in the opposite direction with your other hand. This will start a circular motion in your opponent's upper body.

Push-pull principle in reverse takedown.

2. Your opponent's balance can be taken by stepping with one foot behind your opponent's foot. He is now unable to reposition his C.G. enough to keep from going over your hip. The weight of his upper body will create the momentum.

3. If your opponent reaches out to grab you, try intercepting his arm with a wrist grab. Pull his arm toward you to unbalance him. Pushing with your free hand against his opposite shoulder will increase the dynamics of the technique. Once you have pulled your opponent to close range, step behind his leg until you are hip to hip and can proceed with the throw.

4. In order for a takedown to be speedy and controlled, you must allow your body to lead. Although the circular movement should be synchronized with your feet, body, and hands working together, starting the circle with your feet will usually result in a struggle, with the heavier body trying to catch up. It is better to allow the rotation to start in your body.

Explosiveness and closing distance:

1. It is often the takedown itself that is taught, rather than how to get into position to execute the technique in the first place. A distraction, like a kick, can help you close distance.

2. Throws and takedowns can be initiated from a distance with a tackle. Your gap closure must be quick and explosive, and preferably set up with a distracting move.

118   Fighting Science

A kick can be used to set up a leg takedown.

3. When tackling your opponent, your body must be lower than his, with your focus down and not straight back. The tackle should happen in one fluid motion. Baby steps, or fidgeting, will disrupt the momentum and may give the technique away.

Finding the right distance:

1. Many fights originate with punches and escalate to takedowns and grappling. It is therefore important to develop distance awareness. Try to remain outside of your opponent's reach, yet close enough to be able to counter-strike quickly.

2. There is only one distance that is optimal for a takedown or throw, but a lot of takedowns and throws "end up" at other distances as well. The optimal distance in a throw is touching. If you are too far away, the throw will employ a larger circle and may lose much of its momentum.

3. Once you have closed distance and positioned for the throw, the throw itself must be dynamic and happen without any stuttering or stop in momentum.

4. A takedown may employ a larger circle than a throw. A common mistake is to straighten your arms once the circular motion begins. This will result in stalling the momentum by splitting it into two different directions.

Momentum 119

Momentum and inertia:

1. Your momentum is a combination of body weight and speed. Pulling or pushing your opponent, while simultaneously continuing your own momentum, enables you to launch a takedown or throw.

2. Moving your opponent back, while you move forward, is the easiest way to unbalance him through momentum. You can also rely on his inertia (resistance to change) by suddenly reversing direction.

Pulling your opponent forward, while you step back, gets his momentum going (left). If you suddenly reverse direction and do a leg sweep, his inertia will aid in the takedown (right).

3. Also try circular momentum in conjunction with straight momentum. First, pull your opponent forward, while you step back. Allow his forward motion to continue, while you turn slightly sideways to start the circular momentum of your hip.

Pull your opponent forward to start his momentum. Then execute a quick hip throw through circular motion. Because of your opponent's inertia, it is difficult for him to stop his forward momentum and counteract your throw.

120 Fighting Science

4. Shoot at your opponent's lower legs and pull his legs tight together to restrict his lower body momentum. This will keep him from taking a step for balance.

In a leg takedown, execute a quick shoot and pull your opponent's legs together. His upper body momentum is allowed to continue, while his lower body momentum is restricted.

Momentum when grappling

When your opponent ends up on the ground, you have the choice of continuing the fight, or trying to get away. In a competition environment, you must know how to continue successfully on the ground. The concept of *force per square inch* can be used to pin your opponent's head with your knee. This may be especially effective after a throw. When he can't move his head, he has little opportunity to fight back.

If both of you are on the ground, momentum and explosiveness should be used to reverse positions, allowing you to gain superiority. Momentum can also be used preparatory to a full grappling match.

Circular momentum and balance:

1. In the intermediate fight between stand-up and grappling, as when you are on your knees, momentum can be used much the same way as when standing up. Circular motion, using your knees as pivot points, can now be used effectively.

2. Just as when standing up and attempting a throw or takedown, your move on the ground must be explosive. If the technique is initiated from a distance, you can use your body weight coupled with speed to unbalance your opponent.

Momentum

Inertia and splitting the focus:

1. When grappling with a heavier opponent, try using pain as a distraction. When he reacts, explode with a move intended to allow you to escape or reverse positions.

2. A sudden change in direction creates a moment of surprise, and may enable you to take advantage of your opponent's inertia. If your opponent is fighting your grip, suddenly switching to a different move will make him "lag behind" both physically and mentally.

A figure four lock-up, with opponent resisting, can be reversed to a full armbar, relying on your opponent's inertia (resistance to change).

Reversing positions:

1. When working with a bigger or stronger opponent, keeping your body as tight to his as possible allows you to take advantage of his momentum whenever he is moving. This is true regardless of who initiates the movement. If your opponent is straddling you, and you are successful at shifting his C.G., once he starts losing balance, you can rely on his momentum to pull you into the top position.

2. It may seem easier to throw your opponent off when his center of gravity is high. But when you keep him close, he becomes "one with you," and there is less rotational inertia to overcome.

Mass and momentum quiz

1. What is the difference between mass and weight?

Mass is a measure of the amount of matter in an object. Weight is how heavy an object is when acted upon by gravity. You can be weightless, but you cannot be massless.

2. Name three ways in which you can place the weight of your body behind your strikes.

A. Take a step forward with your strike.
B. Pivot your foot, hip, and upper body in the direction of the strike.
C. Fight from a sideways stance instead of a square stance (horse stance).

3. When throwing a hook, why should the elbow of the hooking arm stay in front of the body?

To generate power in your strikes, you should rely on body rotation. The body is heavier than the fist and arm, and the move should therefore originate in the body. If the elbow of the punching arm stays behind the body, the elbow must catch up, and you can therefore not utilize the full weight of the body for power.

4. When is it easiest to utilize body rotation and why: when your hook is a wild swing (sometimes called the "haymaker" or "John Wayne" punch) or when your elbow stays tight to your body?

Any time anything "sticks out" from the center of your body, you must overcome *rotational inertia* (more about this in Chapter 5). When executing a technique which employs circular motion (hook, round house kick, spinning back kick), you should keep all body parts as close to your center as possible. This will enable you to accelerate the spin, so that the strike, when released, will happen at its maximum speed. Speed, in turn, translates into power.

Study a video of boxer Mike Tyson. Mike Tyson has the ability to create extreme power in his hooks, because of the way he throws his whole body into the technique. When throwing the hook, try to keep your arm very tight to your body. Increase the momentum by exploding with a short jump.

5. Once you have started the motion of a strike, it is important to keep it going. Why?

Any time you change the state of motion (from rest to moving, from moving to rest, or slowing down or speeding up), you must overcome inertia. This takes energy and is destructive to power.

6. What is the difference between speed and velocity?

Speed simply tells how fast something is going. Velocity also has *direction*. If the power of a strike is allowed to split into two or more directions, maximum power in the *intended* direction cannot be achieved.

7. If you are less massive than your opponent, can you still be as powerful?

You can increase power by throwing your strikes at a greater speed. This concept will become especially important when we start learning about *kinetic energy* in Chapter 8.

8. How can you use your opponent's momentum against him?

Time your strikes so that your opponent steps into them. Or grab your opponent and pull him forward and into the strike.

9. Name four ways in which you can speed up your strikes?

A. Use striking that is economical. Avoid any wasted motion.
B. Once you have started a combination, it takes less energy to keep it going than to stop and then restart a second combination.
C. Avoid falling into your opponent's rhythm. Be aware of both your own and your opponent's rhythm and force yourself to stay one beat ahead.
D. Learn to relax.

Glossary

Adding Momentums--When your opponent steps into your strike, your momentums are added. This makes fighting economical, and can make you appear stronger or more powerful than you are.

Gravity--An attraction between objects due to mass.

Impulse--The change in momentum. The faster you can change the momentum, the greater the impulse, and the greater the power.

Inertia--Resistance to change in motion. An object at rest tends to stay at rest; an object in motion tends to stay in motion.

Kinetic Energy--Half the mass times the velocity squared. Kinetic energy depends on the mass and the speed of the object. If a fighter can double his speed, he can quadruple his kinetic energy. Kinetic energy has a great capability of doing damage.

Mass--The quantity of matter in an object. When acted upon by gravity, we can use mass interchangeably with weight.

Momentum--The product of the mass of an object and its velocity. The heavier the object, and the faster it travels, the greater the momentum.

Force Per Square Inch--The narrower the base of an object, the greater the force per square inch on the opposing surface.

Resultant--The sum of all vectors. When throwing a strike, we should strive to make the resultant as great as possible.

Rotational Inertia--Resistance to change in an object that is rotating. It takes a force to change the state or direction of rotation.

Sensory Overload--By striking and kicking with explosive combinations, and to as many targets as possible in the shortest amount of time, you will create "sensory overload" and send your opponent into a state of confusion and chaos.

Vector--An arrow symbolizing the strength and direction of a force. The longer the arrow, the stronger the force. For maximum power, all vectors must point in the direction of your strike.

Velocity--A measure of the speed of an object and its direction.

Direction
F = ma, An Unbeatable Combination

When stepping on the gas pedal in your car, you will *accelerate*. If you were a passenger riding in this car, and you had your eyes closed, you would still know that acceleration was taking place because you would suddenly lurch toward the rear of the car. Acceleration is usually associated with building speed, but in physics it also applies to a decrease in speed as well as changes in direction. If somebody pulled out in front of you, and you stepped on the brakes to avoid a collision, you would now lurch forward instead. Likewise, when you drive up the clover leaf on-ramp to the highway, you will feel your body slide to the outer part of the curve. As mentioned earlier, speed is a measure of how fast you are going, while *velocity* is a measure of how fast you are going as well as which *direction* you are moving in. *Acceleration* is defined as *the rate at which velocity changes*.

In martial arts, the faster you can accelerate *in the direction of your punch or kick*, the more power you will produce. This can be demonstrated through the equation *F=ma*, where **F** is the *force*, **m** is the *mass*, and **a** is the *acceleration*. We have already talked about how a fighter can produce power by placing the weight of his body behind the strike. We have also talked about the importance of speed, and how a fast strike has more power than a slow strike. If you can accelerate a strike with the moment of impact occurring at the highest speed, the force (power) will be even greater. We can also say that a given force divided by a small mass produces a large acceleration. For simplicity, let's say that your 3 has a force of 1000 and a mass of 10. What is the acceleration? We can use the equation F=ma: **1000 = 10 X acceleration**, or we can say that the acceleration = 1000/10, which is 100. The same force divided by a large mass produces a smaller acceleration: **1000 = 100 X acceleration**, or the acceleration = 1000/100, which is 10. If *both* the mass and the acceleration are large, the force will be even greater. For example, a mass of **100** and an acceleration of **100** is equal to a force of **10000**. We can see now why power seems to increase with weight and speed, and why full contact fighting, like boxing and kick-boxing, utilize weight classes.

But will the best lightweight fighter in the world really be defeated every time by an unranked heavyweight in a bar brawl? What is the smaller fighter to do when the fight isn't governed by rules and restrictions in weight?

We have talked about the benefits of mass, and how a massive object has a lot of momentum and is difficult to stop. But a massive object also has a lot of *inertia* (resistance to change in motion). It is more difficult to stop the motion of a heavy fighter than that of a lightweight, but it is also more difficult to start or accelerate the motion of a heavy fighter. This is why lightweights are often faster than heavyweights. Watch boxing, and you will see for yourself.

Speed translates into power. The faster the speed, and the faster the acceleration, the greater the power. So, you see, there is a trade-off. A lightweight fighter can produce the same amount of power as a heavyweight, but his punches and kicks must be faster to make up for the lack of sufficient mass. Later, when we get into *kinetic energy*, we will see how speed can be even more devastating than weight.

Speed in combinations

Because of his mass, a big and strong fighter can be methodical and throw strikes at a fairly slow pace. But a smaller fighter must rely on speed to create power. When throwing combinations, you should work on decreasing the beat between punches and on ending the combination with a "power strike," usually a rear strike where you can build *momentum* through distance. If you get tired and need to rest, then rest between combinations.

The higher the speed of a strike, the more power it will generate. You should therefore try to accelerate your strikes with the moment of impact coming at the highest speed. A combination that involves strikes in different directions is difficult to build speed in, because every time you start or stop or change an object's state of motion, you must overcome inertia. This is why it is more difficult to throw a jab followed by a lead hook off that same hand, than to throw a jab followed by a rear cross with the other hand. Both the jab and the rear cross go in the same direction, while the jab and the hook have perpendicular paths.

The only way you can change direction without stopping and restarting again is through *circular motion*. Because a combination does not consist of strikes only, if you can think of *every* move as part of the combination, then you can also combine blocking, slipping, and bobbing and weaving with your offense. Some of these moves are by nature circular, like the bob and weave. You can now utilize the circular motion of this defensive move to set for offense without stopping motion.

Let's talk about the uppercut. This strike can be thrown in two ways: *separated* from the move that precedes it, or *within* the move that precedes it. For example, a bob and weave to the left followed by a left uppercut must be broken down into two moves: First, bob and weave to the left (the body has now set for the uppercut). Then, reverse direction back toward the right along the same path, and throw the left uppercut within the movement of the bob and weave. The move will look like the lower half of a circle, or a smiley face.

Note: A *bob and weave* is not the same as a *slip and weave*. Bobbing is the vertical motion of your body and weaving is the horizontal motion. A bob and weave is therefore a combination of vertical and horizontal motion; hence the smiley face.

You can also throw the uppercut in one continuous motion within the bob and weave, and build speed throughout. This is done by initiating the move with a slip to the left. You have now moved your head off the attack line along which your opponent's strikes are thrown. Without stopping the motion, continue into a bob and weave back to your right, and throw the left uppercut.

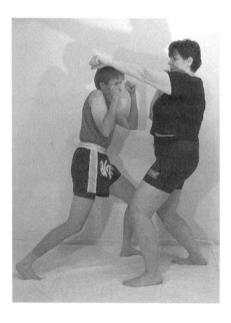 

Martina slips a jab to the left. She weaves back to the right and throws the left uppercut within the motion of the bob and weave.

As long as there are no opposing movements in body mechanics, and your strikes naturally follow one another, it will take more energy to stop and then restart a combination than to keep going. Try the following exercises when practicing punch combinations:

- Evaluate which strikes naturally follow others. Generally, strikes that feel good, in which you are in a balanced stance, and where there are no opposing movements in body mechanics will be powerful. Alternating strikes are some of the easiest to throw because your body will automatically "set" for the next strike. For example, throw a straight punch with your lead hand, followed by a straight punch with your rear hand.

- Work on the "economy of motion" principle, where one strike helps the other. Think of it as a tram going up a mountain. Both the gondola at the bottom and the one at the top start their motion simultaneously, and pass each other at the halfway mark. To duplicate this when punching, first throw your lead strike. As soon as it starts on its way back to the guard position, your rear strike should start on its way out.

- Two strikes that employ different directions are the most difficult to find a natural flow in, especially if both strikes are thrown with the same hand. An example would be a straight lead hand punch followed by a lead hand hook. The first punch follows a straight path forward, while the second punch is thrown at an angle perpendicular to the first. Attaining power in this combination requires a quick move with your body to reset the straight strike, before it can again start in the motion of the hook. This combination also requires more energy because you must completely stop the first strike before your hand can change direction.

The next combination that Martina Sprague demonstrates on the double end bag is one of the more difficult combinations to throw with crispness. To be successful, you must be careful not to split the *resultant* into two or more *vectors* (more about resultants and vectors in the next section of this chapter). Each strike has to come to a complete stop before the next strike can be started. The reason is that the jab, the lead hook, and the lead uppercut are all thrown with the same hand and in three different directions.

When going from the jab to the hook, and from the hook to the uppercut, if the motion is not completely stopped before starting the next strike, you will automatically employ circular motion. The benefit of circular motion is that it allows your strikes to continuously build speed without having to overcome inertia at every start and stop. The danger of circular motion, however, is that unless the strike is thrown absolutely straight on impact, maximum power cannot be attained. *Don't circle your hand at the moment of impact!*

Martina throws three different lead hand strikes, each time employing a different direction.

For power, your body weight must be directly behind each strike (forward with the jab, to the right with the left hook, and upward with the uppercut). If you fail to reset your body's balance between each strike, the next strike will be thrown using arm power only. If you are too hurried, the next strike may not impact the target straight, and the resultant force will split.

- Relaxation increases speed, and increased speed means increased power. Grab a set of light handheld weights--five pounds each, for example--and alternate left and right punches in the air as fast as you can for three two-minute rounds, with one minute rest between rounds. To get the most benefit from this exercise, you must bring your hand all the way back to your shoulder after each punch. Once your arms get a little tired, you are less likely to tense when punching. Be careful not to hyperextend the elbow by snapping the strike on its way out.

- Put your gloves on and punch the heavy bag with straight left and right punches for three one-minute rounds, with one minute rest between rounds. Focus on relaxation by striking as fast as you can and without stopping. When your arms begin to hurt, do not give in to your desire to slow down. This exercise is *supposed* to be tiring.

When punching in combinations, your speed and power should build throughout the combination. This allows you to end with a strong punch. Most fighters find it easier to throw a multiple punch combination than a multiple *kick* combination. This is because our hands are naturally faster and more precise than our feet. Because of this, fighters tend to throw their kicks isolated. They throw a kick at the beginning of a punch combination, and then pause briefly before beginning to punch. Or they throw a punch combination, then pause and throw a single kick, then pause again and throw another punch combination. Many fighters don't throw multiple kicks at all, but rely solely on single kicks. The drawback of this is that the movement must be stopped and restarted many times, which takes energy and is likely to tire the fighter. In addition, the kicks will often be telegraphed because they are not blended in with the rest of the technique. Try the following exercises when practicing multiple kicks:

- Have a partner hold a kicking shield. Throw ten round house kicks in rapid succession with your lead leg. Your foot must come to a complete stop between each kick before reversing direction. This exercise takes energy and requires relaxation for speed.

- Practice multiple kicks with both legs on the heavy bag. Do not attempt to stop the swinging bag, but adjust to the angle and distance by taking tiny steps around the bag. Throwing multiple kicks effectively requires very good timing, coordination, mechanics, and speed.

  Note: There is a difference between throwing a multiple kick *combination*, and simply throwing many single kicks.

- Start the combination with a punch and evaluate which type of kick would most naturally follow. Because your legs are longer than your arms, unless your opponent steps back to give you distance, some kicks, like the side kick or spinning back kick, may feel crowded if thrown after a hand combination.

- The round house kick can be thought of as a universal kick, because it can be thrown with ease at the beginning, middle, or end of almost any punch combination. This is because the round house kick follows a curved path and does not, in the same way as other kicks, rely on distance between you and your opponent. If you are at long range, you may connect with the instep or ball of your foot. If distance is middle range, then connect with your shin. If you are very close, you may convert the round house kick to a round house *knee* strike.

The resultant force

Power relies not only on force, but also on *direction*. To obtain maximum power, we must ensure that the sum of all vectors is as great as possible and in the direction we wish to throw the strike. A *vector* is the combination of the *magnitude* (strength) of the force and its *direction*, and is symbolized by an arrow. ⟶

When throwing a punch which is focused straight forward, and with all of your body weight behind it for power, the resultant force will be in the direction of the punch. But if you throw a punch while simultaneously taking a step back, the punch will lose much of its power because the resultant will split, with one vector going forward and one back.

Think of it this way: A vector is illustrated by an arrow pointing in the direction of the force, with the length of the arrow representing the strength of the force. When you throw a punch, the vector can be thought of as being parallel with your arm and in the direction of the punch.

The vector is parallel with the fighter's arm and in the direction of the punch.

To calculate the power, all vectors in the same direction must be added, with those in the opposite direction subtracted. The final result of this is called the *resultant*. If the punch is thrown absolutely straight with the full weight of the body behind it, the resultant will be at its maximum. If the fighter takes a step back while simultaneously throwing a punch forward, we will have two vectors in opposite directions. When we add these two vectors (subtract the rearward one from the forward one), the resultant force will be less than in the first example, and the punch will not have as much power. When two or more vectors are going in different directions, the force will *split* because the vector components are not lined up with the resultant. Maximum power can therefore not be attained. A few examples would be:

- Leaning back excessively when kicking. Only lean back enough to keep your balance.

- Looping a punch. A jab or rear cross should come straight out from the guard position. A common mistake is to raise the elbow prior to punching. In addition to not allowing you to keep the full weight of your body behind the strike, this also creates a vector toward your centerline. The force of the vector striking the target will now be shorter than the resultant, and maximum power cannot be attained. Another common mistake is to loop the punch slightly downward when retrieving the fist after impact.

- The spinning back kick is especially interesting because of its combination of circular and linear motion. When initiating the kick, the spin should be used to accelerate it. But when the kick connects, your leg must be linear toward the target. Throwing a powerful spinning back kick requires target accuracy and a feel for exactly when to convert the circular motion to linear.

Caution: Don't confuse the spinning back kick with the spinning *heel* kick, which is designed to strike its target from the side.

Keith initiates the spin in his upper body.

The circular motion is then converted to linear, with the kick thrown straight.

- Stepping side kick. Some martial artists are taught the crossover side kick, where you initiate the forward motion of the kick by moving your rear foot behind your lead foot, so that your legs are crossed. Stepping will close distance and create momentum, but crossing your feet may redirect the path of the kick slightly, setting your hips at a different angle so that the power is no longer projected straight through the target. For fastest gap closure, your step should be as "clean" as possible. I recommend taking a half-step only, by bringing your rear foot forward half the distance to your lead foot. This decreases the risk of telegraphing the kick, or of narrowing your base too much (loss of balance), and of splitting the power into two or more vectors.

- The front kick, when thrown *vertically straight*, will often allow your foot to slide against your opponent instead of penetrating him. A good way to practice penetrating force with the front kick is on a heavy bag that is held steady by a partner. Keeping the bag vertically straight will force you to throw the front kick at a slight *diagonal angle* upward and into your target. There is an exception to this: if your opponent is bending forward, you may throw a front kick

136 Fighting Science

that is vertically straight to your opponent's chin. The kick will now miss all of your opponent's body and hit his chin only.

In a takedown, it is equally important to focus all energy in the direction you want your opponent to go. If you want him to go down, you must focus all energy toward the ground. Additional power can be attained by keeping the technique close to your body, and by using your own weight to aid in the takedown.

Martina drops her opponent to the ground by applying pressure against the wrist. Note how the technique is kept close to her own center of mass, allowing her to use her weight to aid in the takedown. Note also how her focus is toward the ground. You can see this by the slight hunching in her upper body.

The dangers of exhausting the motion

Momentum can be built by eliminating the start/stop movement. When you stop the motion of a technique, you must expend energy to restart it. A technique that requires no start/stop is therefore more economical than one that does. In the pictures above, Cindy throws a left hook, followed by a left kick to the leg. Because the kick naturally follows in the same motion as the hook, there is no need to reset the body between strikes, and the speed of the technique can be increased. This is contrary to a front kick followed by a side kick, which require different turns in the hips, with a slight pause between kicks to reset the body for the next technique.

There are, however, some drawbacks of the technique pictured above. Let's look at some points that must be considered for the effectiveness of the technique:

1. If you allow the hook to exhaust its motion before you start the kick, the kick will be almost powerless. This can be likened to taking a deep breath and exhaling until all the air is gone from your lungs. Now, in order to exhale again, you must first inhale. If the motion of the hook has exhausted itself, you must first pull back and chamber for the kick in order to attain power. This defeats the purpose of flowing from one technique to the next. You must therefore start the kick *before* the motion of the hook has been taken to completion. This is also one reason why alternating strikes (left/right) seem more natural than strikes thrown with the same hand/foot; your body will automatically chamber for the follow-up strike. However, starting the kick before the hook has gone through the target, may not give you enough time to determine if there is a legitimate target for the kick. The kick may be wasted motion, and may even set you at a disadvantaged position. Because a fight is dynamic, the skillful fighter doesn't plan too many moves in advance, but waits to see how the fight unfolds, and then adapts within.

2. The second concern is your opponent's counter-strike. It is unrealistic to think that

you can bombard your opponent with strikes without expecting a response. Because the shortest distance between two points is a straight line, a straight punch will always beat a looping punch. A dangerous counter-strike that you must look for is therefore the rear cross. If your opponent blocks your hook with a forearm block, he is likely to attempt a straight strike to your facial area after the block. He can also use the block to "freeze" you for a fraction of a second, and taking command of the fight.

3. This technique has a serious catch-22. If you have started the round house kick (which you must if you don't want to exhaust the motion of the hook) when your opponent throws the counter-strike, you will be in a disadvantaged position for absorbing the power of the strike. You will now have a narrow base on one foot, with full focus of power into the kick. The effects of your opponent's counter-strike will be more adverse than if you were in a stable stance and not kicking.

The overhand projectile

A strike that seems to end many full-contact matches is the *overhand strike*. The advantage of this strike is that it can be thrown from a tight distance where you would normally throw a hook or an uppercut. The strike is therefore seldom expected. As your opponent covers in preparation for a hook or uppercut, the overhand strike will find an open target not from the side or below, but from over the top. Because of the downward path of the strike, it also allows you to drop your weight and use gravity to your advantage.

Force = mass X acceleration
or
Force = ma

Force = m**a**

As demonstrated by Keith, the overhand strike comes over the top, allowing you to use the force of gravity to accelerate the strike on its downward path.

Caution: The overhand strike employs circular motion, and we must therefore be careful with the resultant force. A common mistake is to pull the fist in toward your body and split the force into two or more vectors.

The overhand strike can be thought of as a *projectile*, following a curved path, and will accelerate on its downward motion. More acceleration means more power. The target should therefore be struck *after* the overhand strike has reached maximum height. The curved path of the overhand strike has two advantages:

- It allows you to use the force of gravity, where the vertical vector component is longer than the horizontal vector component on the downward motion of the strike.

- The overhand strike can be thrown from a tight distance, where a straight right would not be possible.

A projectile accelerates only in the vertical direction, while moving at a constant horizontal velocity. In the case that follows, the vertical vector component is longest at the beginning and end of the path, with a reduction to zero at the maximum height. The horizontal vector component is the same everywhere. The actual velocity, and therefore the power, is represented by the *resultant*; the *diagonal* vector.

140 Fighting Science

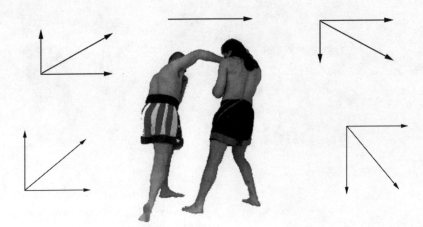

The overhand strike should land when the resultant is at its longest. Gravity by itself may not add significant power to the strike at that short a distance, but power can be further increased by dropping your weight simultaneous with the punch. In addition, when a fighter gets tired, the benefits (or disadvantages) of the force of gravity can be felt more readily.

Summary and review

It is generally understood that a person's weight and speed is combined to form momentum. But less often do we think about, or fully understand, the significance of direction. In fact, this is one reason why the inexperienced martial artist often fails to produce a powerful strike or kick; because he allows the opposing force to set him back. He fails to commit fully to the strike, stopping short or holding back on impact.

Direction when striking

Direction when striking can be summarized into three principles:

- Throw combinations that naturally flow together.
- Avoid splitting the resultant by relying on proper body mechanics.
- Utilize circular moves to accelerate your strikes.

Natural combinations:

1. A natural combination allows you to flow from one punch or kick to the next. Alternating strikes in the same direction generally flow together naturally.

A lead jab followed by a rear palm strike is an example of a natural combination.

2. As long as the strikes are thrown with alternating sides of your body, a natural combination can also employ two different directions.

A lead jab followed by a rear uppercut is an example of a natural combination employing different directions.

3. Circular motion allows the second strike to follow naturally off the first, with the added benefit of a build-up in speed.

An inward hammer fist followed by a spinning back fist is a natural combination that allows you to build speed through circular momentum.

Splitting the resultant:

1. The resultant is the sum of all vectors, along which the force of a strike is the strongest. The resultant depends on the direction of the strike.

2. When the resultant is split, the force of the strike cannot be focused into a single point. The strike is therefore deprived of power.

Pawing when punching automatically splits the resultant force (left). This is why proper body mechanics is so important (right).

Acceleration and eliminating the start/stop:

1. Think of every move as part of the final outcome of a technique. Choose moves that allow you to flow smoothly from one technique to the next. This decreases the need to slow down or stop. Circular motion allows you to switch direction without stopping.

Providing that the direction of the moves don't oppose one another, the combination of a hook, a bob and weave, a step, and a round house kick, has the potential to accelerate without the necessity to stop the motion.

2. Try to accelerate the strike so that the fastest speed happens at the moment of impact. Because your body is stronger than your arm, the movement should originate in your body.

Direction when kicking

Kicks rely on the same principles of direction as strikes. If attacked by more than one person, and your opponents come from different directions, it is almost impossible to focus your power fully. Kicking an opponent behind you, while simultaneously striking an opponent in front of you, are moves that employ opposite directions, and will only be successful if you can time it so that your opponent walks into your kick. If two opponents approach from the sides, and you strike with both hands simultaneously, you will again have a problem with the direction of energy.

144 Fighting Science

Natural combinations:

1. Natural combination kicks all flow in the same direction. There must be no stop in momentum. Generally, kicks with opposite legs classify as natural.

2. Pay attention to body mechanics. It is possible for two kicks to be thrown along the same path, but because the direction of the body differs, they don't classify as a natural combination.

The front kick and side kick with the same leg do not classify as a natural combination, even though they are both thrown in the same direction. This is because the hips are turned forward with the front kick (left), but at an angle perpendicular (or nearly perpendicular) with the side kick (right).

Splitting the resultant:

1. The resultant will split any time there are opposing movements in body mechanics. Perhaps the most common is leaning in the opposite direction of the kick. Some incline in the upper body may be required in order to maintain balance. Try crouching slightly, rather than leaning. This will help you project the power in the direction of the kick.

Leaning when kicking splits the resultant force (left). Crouching helps you project the force forward (right).

2. You also risk splitting the resultant if the kick lacks full penetration. This is commonly seen when kicking a target that is much heavier than the practitioner. Because of the psychological factor when confronted with a heavy opponent or bag, we tend to allow the target to knock us back, rather than using our kick to knock the target back.

3. When stepping in with a kick, the momentum must be allowed to continue past the target. If you break the momentum before letting the kick go, much energy is expended trying to slow yourself down.

Acceleration and eliminating the start/stop:

1. Acceleration in kicks is best built through distance. Because a circle covers a longer distance than a straight line, spinning kicks can be accelerated easier than straight kicks. Try combining moves that flow naturally with the spin.

An inward forearm block followed by a spinning back fist have the potential to reach great speed and power.

2. If your opponent sets you spinning, try going with the motion to accelerate your counter-attack.

An attempted sweep can set your body spinning and build power for the spinning back fist.

Direction in defense

As you will see, the same principles that apply to punching and kicking also apply to defense. But you should think one step further than avoidance. Much of defense is about keeping your opponent from using the principles of physics to his advantage. I'm sure that you have seen a fighter beat the heavy bag with what appeared to be awesome power. And if you were up against this fighter next, you would probably be intimidated by his strikes. But when we get to the actual fight, our strikes don't seem to work as well. This is because bags lack the ability to defend. Your defending block or parry will split the resultant force in your opponent's strike (if ever so little), and the strike will not be nearly as damaging at it appeared on the bag. This is also why a person can't learn to fight by beating bags alone; it is not just your ability to throw a strike mechanically correct that matters, but also your opponent's ability to move, defend, and counter the attack.

Natural combinations:

1. When you have finished a move, you must reset your body's balance. For example, if you place more weight on your rear foot, you must eventually move your upper body forward again to its balanced state. If you duck a strike, you must eventually come back up again. If you move to the side, how can you use the momentum to trigger offense?

2. An example of a natural defensive combination is the bob and weave with an embedded uppercut, or a step to the side to avoid a strike, with a round house kick embedded in the lateral step.

Splitting the resultant:

1. The resultant force will split any time you use opposing movements in body mechanics. When blocking and kicking at the same time, care should be taken to use a block and kick that require the same turn in the body.

Acceleration and eliminating the start/stop:

1. Any technique that allows you to use defense to accelerate offense will increase the power and speed up your counter-strike. An example is the "pick and counter." This move helps you launch offense by utilizing a small circle with the defending hand.

Direction 147

When parrying, note how the downward circle with the defending hand draws the counter-strike out. This is an example of the push-pull principle, and utilizing both sides of the body for efficiency.

Direction in throws and takedowns

Momentum in throws or takedowns is achieved by keeping the motion continuous and accelerated through a circle. Direction can be a bit tricky, because you usually have more than one direction to contend with. Common errors when attempting to throw an opponent include pushing your opponent back, but neglecting to push down, or pushing down, but neglecting to circle.

Natural combinations:

1. Natural movements in takedowns and throws involve using your opponent's weight and energy to your advantage. Once he starts going over your hip, allow him to fall. This keeps you from carrying all the weight, and will make the throw seem almost effortless.

2. If a stronger opponent resists a joint lock preparatory to a takedown, your ability to flow with the motion into another technique may determine how successful you are.

148 Fighting Science

Splitting the resultant:

1. Sometimes, splitting the resultant force purposely may allow you to use your opponent's inertia against him. When you pull your opponent off balance, he will have a tendency to resist. By reversing direction quickly, you can upset his balance.

2. If you attempt a joint lock preparatory to a takedown or throw, and your opponent resists, try going the opposite direction. In order to use his inertia against him, the reversal of direction must happen quickly.

Martina attempts an outside wrist lock (left). Keith resists (middle). Martina reverses direction to an inside wrist lock (right).

Acceleration and eliminating the start/stop:

1. In a throw or takedown, you want to keep your own body as stationary as possible. This is because the smallest movement is in the center of the circle. A lot of movement in your opponent's body will unbalance him easier.

2. When using circular motion in a takedown, think of the axis of rotation as extending vertically through your body. If you take many small steps, you risk pulling your opponent in a circle, rather than taking him down.

Stepping excessively will result in your opponent maintaining balance, and will lessen the effect of the takedown.

3. If you have to close distance from long range, try to initiate the circle as you step in. This will keep the momentum going, and will help accelerate the technique.

Direction when grappling

When on your hands or knees, or on your side or back, the pivot points will change to whatever area of your body you are supporting yourself on. You must now learn to think in a slightly different dimension, but still within the same principles of direction. When on the ground, your body weight should be used to achieve dynamics in the proper direction.

Natural combinations:

1. Try to think of yourself and your opponent as one entity. This will enable you to flow with your opponent's motion and use his body weight to your advantage.

2. A joint lock is only damaging because it goes against the natural movement of the joint. If your opponent has applied a joint lock, going with the motion of the technique may enable you to break free.

The twist lock against the wrist can be escaped by going with the motion of the technique into a forward roll.

Splitting the resultant:

1. A sudden change in direction on the ground allows you to use your opponent's inertia against him.

2. If your opponent is straddling you, and you are successful at shifting his center of gravity but fail to use your body weight in the same direction, your attempt to reverse positions will come to a halt.

When your opponent is straddling you, pushing up on his chin to shift his center of gravity, but failing to raise your hips, will split the resultant, because the weight of your body acts in the opposite direction of the force.

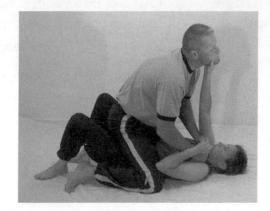

Acceleration and eliminating the start/stop:

1. Whenever your opponent initiates a move, try to accelerate it with circular motion.

2. Look for times when your opponent is slightly off balance. You can now use his body weight against him.

When your opponent reaches forward, grab his arm and pull him off balance. Accelerate by using your body weight in the same direction.

Inertia and vector quiz

1. Acceleration does not refer only to speeding up. Explain.

Acceleration is defined as the rate at which velocity changes. But since velocity is defined as the speed *and* the direction, acceleration is taking place whenever you are speeding up, slowing down, or changing direction (as in circular motion). As martial artists, we are especially concerned with the direction of our strikes. Any "veering off" will split the resultant force into separate vectors.

2. What is meant by a natural combination?

A natural combination is easy to throw with no awkward movements between strikes. A natural combination has the ability to build speed quickly, because there is no stopping and restarting of motion. In general, natural combinations rely on alternating strikes in the same direction. For example, a jab followed by a rear cross, or a jab followed by a rear leg front kick.

3. Why is it more beneficial to throw one lengthy combination than many single strikes?

Many single strikes require a constant starting and stopping and restarting of motion (overcoming inertia). A lengthy combination allows you to build speed (which is important to power), with the highest speed occurring at the end of the combination.

4. Explain why distance is less critical when throwing the round house kick than when throwing a front kick or side kick.

Because the front kick and side kick are straight kicks, they rely on the footage between yourself and your opponent. If you are too far away, the kick will not land. If you are too close, you will jam your own kick and stifle power. The round house kick, on the other hand, employs slightly circular motion and is intended to strike your opponent from the side. Impact can be made with your instep, shin, or knee, depending on how far away you are.

5. Name four mistakes in body mechanics that have the effect of splitting the resultant into separate vectors.

       A. Leaning back excessively when kicking.
       B. Looping a punch instead of throwing it straight.
       C. Turning your hips too much with the stepping side kick.
       D. Moving back when impacting a target.

6. How do you increase the power of the overhand strike?

The overhand strike follows the path of a projectile, where only the vertical vector component will change in size. The strike should therefore land on its downward motion, where you can drop your weight and use gravity to your advantage.

Glossary

Acceleration--Changes in speed and/or direction. An object that is in motion *and* changes its direction (a car driving up the clover leaf on-ramp to the highway), will accelerate even if there is no change in speed. You can feel that acceleration is taking place by the way your body lurches forward, back, or sideways.

F=ma--Force equals the mass times the acceleration. A lot of weight and a lot of speed will produce a lot of power. A lightweight fighter can make up for the lack of sufficient mass by being faster than a heavyweight.

Force--Any influence that can cause an object to be accelerated.

Inertia--Resistance to change in motion. An object at rest tends to stay at rest; an object in motion tends to stay in motion.

Kinetic Energy--Half the mass times the velocity squared. Kinetic energy depends on the mass and the speed of the object. If a fighter can double his speed, he can quadruple his kinetic energy. Kinetic energy has a great capability of doing damage.

Mass--The quantity of matter in an object. When acted upon by gravity, we can use mass interchangeably with weight.

Momentum--The product of the mass of an object and its velocity. The heavier the object, and the faster it travels, the greater the momentum.

Projectile--An object following a curved path from over the top (overhand strike). In the path of a projectile, only the vertical vector component will increase or decrease in size, while the horizontal vector component remains constant. Because of gravity, the resultant will be at its longest in the downward motion of the projectile.

Resultant--The final sum of all vectors. When throwing a strike, we should strive to make the resultant as great as possible.

Vector--An arrow symbolizing the strength and direction of a force. The longer the arrow, the stronger the force. For maximum power, all vectors must point in the direction of your strike.

Velocity--A measure of the speed of an object and its direction.

Rotational Speed And Friction
Circular Movement = Power

Jump kicks are often used by martial artists in competition to impress the judges and the audience. Some claim that these kicks have no value except for aesthetic reasons. These kind of kicks are often referred to as flashy. Others claim that jump kicks were originally designed for kicking men off their horses. But today when we don't fight against horseback riders, these kicks are of little value other than to show off our athletic ability. True or false?

An object in motion tends to stay in motion unless acted upon by some outside force. I'm sure we have all heard of ***Newton's principles of motion*** some time during our years in school. But if an object in motion tends to stay in motion, why is it that if you roll a ball over a flat surface, it will eventually come to a rest? Why doesn't the ball continue in motion forever? Well, it would, unless *acted upon by some outside force*. The outside force in this case is ***friction***. Friction occurs when two surfaces slide over one another. Less friction means more ease of movement. Different substances have different amount of friction and, if we can reduce the friction as much as possible, our techniques will be faster, more powerful, and thrown with less effort.

Friction is not restricted to solid surfaces, but occurs also in liquids and gases. Friction depends on the nature of the liquid or gas. For example, it is greater in water than in air. Friction in water and air is also less than friction between solid surfaces that slide against one another. But if this is true, then why is there less friction on ice than on pavement? Both are solids, right?

An ice skater can move faster on ice than a person wearing regular shoes, because the blade of the ice skate is very thin, focusing more weight over a smaller surface area. The pressure of this extra weight lowers the melting point of the ice and changes the state from a solid to a liquid. A liquid has less friction than a solid, and ice "appears" slippery because of the thin layer of water between the blade on the skates and the ice. Bet you didn't know that!

But martial artists seldom fight on ice, and even if they did, the slippery low friction surface of the ice hardly seems like an advantage. Having solid ground beneath your feet will give you more friction and better traction, and therefore a lesser risk of losing balance. As we develop our kicking techniques, however, we should study the friction between our bodies and the air through which they move. Friction in air is less than on a solid surface or in water.

Jump kicks

When I was teaching karate, we often did a drill called kicks across the room, where students put together kick combinations that the rest of us had to follow. The youngest student in my class, a little boy about eight years old, always wanted to do the *tornado kick*, which includes both a jump and a spin. This was to the great dismay of the older martial artists who didn't have the same athletic ability. In competition kick-boxing, however, you will mostly see the three basic kicks: the front kick, the round house kick, and the side kick, with an occasional spinning back kick or spinning heel kick. In a full-contact environment, jump kicks are time consuming and may be dangerous to throw if the practitioner doesn't have full control of motion. But when analyzed according to the principles of physics it is found that, aside from getting better height out of the jump kick, a fighter can turn his body more easily in the air. This is because there is less friction between the air and the fighter than between a solid surface and the fighter. Easier turn of the body translates into more speed and therefore more power.

For a jump kick to be powerful, impact should come at the apex, when the jumping foot is at its highest point. If the jumping foot is allowed to replant on the floor before the kicking foot makes contact, the power will split into two directions: *horizontally* into the target, and *vertically* down into the floor. Power can also be increased through momentum by launching your body toward the target. A common mistake is to lean back when jumping. This would have a contradictory effect by splitting the power into two vectors.

When I fought in a kick-boxing match a few years ago, I felt that I was dominating the fight through every round. When I got back to my corner after the last bell, I was certain that I had won. When they announced a split decision in favor of my opponent, I was utterly confused. Later, her trainer came into my dressing room and said that they had ruled in her favor because she had thrown more *high* kicks than I had. But this was a *leg kick fight!* I accepted my loss, but started thinking about how you can increase the power of the low kick even more. A flying high kick, if performed by a person with athletic ability, can have

Rotational Speed and Friction   157

devastating power. But what about a flying *low* kick? Or is there even such a thing?

In full contact kick-boxing, I often rely on the low kick, because it is fast, easy to throw, and takes little energy. I decided that if the jump could be used as a *fake*, you would increase your chances of landing the low kick. Because the jump kick is generally a high kick, once your opponent sees the need to defend against it, he is likely to freeze in preparation for the blow. A jump kick can be faked by looking at the high target when you initiate the jump. When your opponent tenses, you land the kick to his legs. Because there is less friction when your body is allowed to turn in the air, the jump helps you increase the power regardless of whether the kick is thrown high or low. This increases the effectiveness of the low kick, yet eliminates the danger of performing the high kick.

Jumping, sliding, and stepping

The best time to punch or kick is when your opponent is in the process of kicking and is on one leg. There are two reasons for this: First, a fighter is unable to move out of the way when on one leg. Second, balance is greatly diminished when the foundation is narrow (**concept: narrow base = unstable**).

Keith Livingston lands a jump spinning back kick on opponent who is on one leg.

Another good time to kick is when your opponent is moving toward you. You can now use the principle of ***adding momentums***. **Momentum = mass X velocity.** In earth's gravity, mass is interchangeable with weight. Velocity is the same as speed, with the added benefit of *direction*. The impact is greater if struck when stepping toward your opponent than if struck when standing still, because the momentums of both fighters can be added. By timing your strikes right, you can exploit your opponent's strength: his weight and his speed.

The spinning back kick and side kick are two of the most powerful kicks. However, the power of these kicks often fails to mark an opponent for two reasons:

- These kicks require more movement than the front kick or round house kick, allowing the opponent time to move out of range.

- We often have a tendency to move our upper body to the rear, thus splitting the power with some going forward and some backward. Moving your upper body to the rear also decreases your reach, giving you less penetrating force.

Power can be increased by training your body to stay compact. As you start to rotate for the spinning back kick, you should simultaneously lower your center of gravity and stay low until impact is complete. Another benefit of staying compact is that you will expose fewer targets to your opponent's retaliation.

When throwing the jump spinning back kick, the same principle applies. If you raise your body up, you will have a tendency to lean away from the direction of the kick and split the power into two vectors. When kicking a person who is heavier than you, penetrating the target may be difficult. Staying compact enables you to focus all your energy forward and into the target, and use momentum and explosiveness to your advantage.

Because there is less friction between a fighter's body and the air, than between his feet and the ground, he can rotate for the spinning back kick more easily in the air. Since power in the spinning back kick to a great extent relies on how fast the practitioner can spin, less friction means an easier and faster spin. This extra speed also makes it difficult for your opponent to see the kick coming. Even if he sees it, he must first stop his forward momentum, which takes energy, and then reverse direction or side-step to get away from the attack. Varying the shape of the body allows the practitioner to speed up or slow down the kick. A body that is contracted (made small with no limbs sticking out until it is necessary) can spin faster than a body that is spread out.

Rotational Speed and Friction 159

Keith contracts his body (less rotational inertia) before extending the kick. This allows him to increase the speed of the spin.

Note also how the kick itself is executed at the apex (the highest point in the jump). This, too, is done for the benefit of power. If a kick that is designed to go horizontally through the target connects on a fighter's upward or downward motion, the resultant force will split into two or more vectors. In other words, you will have a conflict with the direction of energy, with some energy going into the target, and other (wasted) energy going either upward or downward, depending on whether you throw the kick while gaining height or while coming back down.

Keith executes a jump spinning back kick at the apex of the jump.

A jump round house kick, too, has the potential to be more powerful than a regular round house kick, because it allows the practitioner to rotate his body more freely in the air. The trade-off is, of course, that kicking and jumping require coordination and athletic ability.

Friction always acts in a direction opposing motion. If you are going forward, friction is acting backward, and vice versa. The friction of sliding is somewhat less than the friction that builds up before sliding takes place. That is why it is important not to lock the tires when braking a car. When tires lock, they slide, providing less friction than if they roll to a stop. This concept can be applied to the stepping/sliding side kick. As your kick makes contact with the target, your supporting foot should slide a few inches forward. This provides additional power, reach, and momentum.

To preserve maximum penetration, it is important that distance is not increased by leaning back. Again, your body should stay compact with your center of gravity low. The stepping side kick can be thrown with a small step only to increase explosiveness and conceal movement.

Circular motion

Let's look at **Newton's First Law Of Motion** again: *An object in motion tends to stay in motion unless acted upon by some outside force.* We have already talked about that the tendency of things to resist change in motion is called inertia. The law of inertia is extremely important to us as martial artists, and applies to a wide variety of techniques. For example, a heavy fighter can generate more power than a lightweight, because the heavier fighter possesses more mass. But mass also relates to inertia. The greater the mass, the greater the inertia. The heavier the fighter, the more difficult it will be to stop his forward motion. But the more difficult it will also be to start any motion at all. And without motion, nothing happens. The higher the speed, the more power. If mass helps you increase power, but mass *also* resists the building of speed, one may ask whether it really is to your benefit to be massive in the martial arts.

It takes more energy to change motion, than to simply continue motion that has already been started. It takes more energy to start and then stop and then restart motion, than to simply keep going. Changing the *direction* of motion also takes energy, because we now need to apply some force in order to change direction. Take a lead hook thrown off of a jab. As stated earlier, the jab is thrown straight, and the hook is thrown at an angle perpendicular to the jab. That is why this combination seems more awkward than a jab followed by a rear cross. Going from circular to linear is also more difficult than to simply keep going circular.

Take a look at the following concept:

The shortest distance between two points is a straight line.

If you were to walk from point A to point B and back again, you would start at A, stop at B, then turn around and walk back toward A where you would stop. Or you would start at A, stop at B, and then walk backwards back to A.

In the case illustrated above, you must start, stop, and restart motion in order to get from A to B and back to A. This starting and stopping is not only time consuming; you must also overcome the inertia of motion, which wants you to keep going forward without stopping. Try the following common and easy to throw combination: a left jab followed by a right cross. First, your left jab comes forward. But in order to pull back and start the right cross forward, the jabbing hand must first come to a complete stop. Then the motion of the jab must be restarted in the rearward direction. Because we are trying to build speed for power, this stopping and starting is time and energy consuming. In order to increase the speed and power in our combinations, we need to find a way to change direction easily without having to stop and then restart the motion.

The secret to building speed is in a *circle that never stops*. Whenever you are moving circular, you can change direction constantly without ever having to stop moving. Because of this, we are now going to modify the jab/rear cross combination slightly in order to eliminate the stop/start movement.

 This circle enables you to get from A to B and back to A without having to stop motion. The problem, however, is that the shortest distance between two points is a straight line, so moving in a circle covers a longer distance than moving in a straight line. How do we fix the problem?

162   Fighting Science

This oval enables you to get from A to B and back to A without having to stop motion. The more drawn out the oval is, the less distance you need to cover, and the more economical the strike will be.

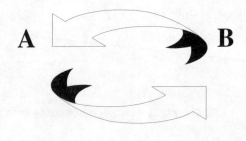

When your jab has landed and it is time to pull that hand back, rather than stopping the motion and then restarting it, try a *tiny* circle in the downward direction and back toward you. The trick is to make the motion *continuous*.

Caution: Making a tiny circle is not the same as "pawing." Be aware of splitting the *resultant* into separate *vectors*. When the strike connects, it must come absolutely straight. If the strike is circular on impact, the power will split with some going into the target and some toward the ground. After the strike has done its damage, however, a tiny circle on its return path will aid in speed when retrieving the punch. The path should not be so circular that your hand drops below the level of the chin, but more of a long, elongated oval. The circle must possess *hairline precision*, so that you don't split the resultant prior to impact.

Rotation of the fist

Power and accuracy of a punch can also be increased by rotating your fist to the horizontal position just prior to impact. Your hand now becomes like a drill, pushing through your opponent's guard and enabling you to take small and well-guarded targets. Because the point in the middle of a rotating system doesn't move, the power is focused over a very small area. This applies to both pointed and blunt objects (your fist). You can test this by attempting to push your fist straight down into a pile of sand. A very large force is needed to move the sand grains to the sides. But if you rely on a drilling motion, the center of your fist will begin to penetrate.

The rotation should start in your foot, hip, and shoulder. It now seems natural to continue this rotation through your arm and fist. A word of caution: If you start the rotation of your fist too soon, you will have a tendency to raise your elbow prior to punching. This will cause you to throw the strike wide without the mass of your body directly behind it, and with power loss as a result. In addition, a strike that is thrown wide is not as deceptive as one that is thrown tight.

Because it is not possible to turn the hand a complete 360 degrees, the drilling effect will work best if impact comes slightly before full extension of the arm. Rotating your fist on impact can also have an abrasive effect, causing cuts on your opponent's face.

Note: Bruce Lee's one-inch punch is thrown with the fist in the vertical position. Because the distance to the target is so short (one inch), twisting the fist to the horizontal position requires slightly more time. The vertical punch gets much of its power from the extension of the elbow and the short explosive move of the body (see "The physics of the one-inch punch" in Chapter 7).

Torque

You are driving your car alone on a desolate road at night, when you get a flat tire. Because you are a person of small build and not very strong, you worry about getting attacked when changing the tire. The tire change must therefore happen quickly. You chock the wheels, jack the car up, and get your wrench from the trunk. But the lug nuts won't budge. You remember when you had the tires rotated last month, and the repair shop used power tools to tighten the lug nuts. You give it everything you have, but you just don't seem to have the strength. You look around. There is no one near, you don't have a cell phone, and there are miles to the nearest building. You have to get this tire changed. What should you do?

Even though you are a trained martial artist, you never get complacent. You know that awareness is the best defense. You look around and find a three foot long pipe in the ditch. "This can be used as a weapon," you think, then get a better idea, remembering what your instructor taught you about *torque*. You grab the pipe, slip it over the wrench handle, and pull. The lug nuts turn. The rest of the tire change is easy, and it's only a matter of minutes before you are moving again, safe now, with the pipe stowed in your trunk for future use.

> **Torque = lever arm X force**
>
> This can also be thought of as the product of the distance from the pivotal point and the force that tends to produce rotation.

Torque, in the martial arts, is often misunderstood and thought of as an impact concept. When I was new in the martial arts, my instructor told me that "torquing" my hand on impact (meaning to turn the hand from the vertical to the horizontal position) would produce a greater force (see "Rotation of the fist" above). But torque does not relate to impact in the same way as impulse or momentum do. Torque is a *leverage concept*, and the idea is not to use more force to increase the torque, but to use a longer lever arm, enabling you to use *less force* to do more damage, and at the same time conserving energy. Speed is not even relevant. As a matter of fact, torque can be increased by inching extremely s-l-o-w-l-y. It's how much leverage you have that matters, and not what your force is on impact.

Torque can be thought of as leverage in the rotational plane. In the picture below, Martina Sprague uses torque against her opponent's elbow joint. The pivotal point is the elbow, the lever arm is the distance from the elbow to the wrist (where Martina's hand is), and the force is the pressure that she exerts by pulling up on the wrist. The more you increase the distance (the lever arm), the more you increase the torque. This allows you to use a lesser force to produce the same amount of torque. If Martina was grabbing her opponent closer to the elbow instead of at the wrist, it would be much more difficult to do damage.

Martina uses torque to control her opponent.

Why is the pivotal point at the elbow and not at the shoulder? Normally, the pivotal point would be at the shoulder, where the arm is attached to the body. But because the elbow is less flexible than the shoulder, it is easier to control your opponent by the elbow. For best results, the force should be applied at a 90 degree angle (perpendicular) to the lever arm. The torque can also be increased by extending the lever arm.

Torque works great in throws. The picture sequence below shows Martina catching her much bigger opponent's round house kick in the crook of her arm. Note the 90 degree angle between Martina's arm and her opponent's leg (the lever arm). Martina now extends her arm at a 90 degree angle to her opponent's leg, and takes him down backwards.

Martina catches Keith's round house kick in the crook of her arm, utilizing torque to throw him off balance.

In order to utilize torque to its fullest, you must catch the kick as far from the pivotal point as possible. The pivotal point is where the leg is attached to the body. It is therefore better to catch the kick close to your opponent's foot rather than close to his hip. Remember, the longer the lever arm (the distance from the pivotal point to the force), the less force you will need for a given amount of torque. If the kick had been caught at the knee joint instead of at the ankle, more force would be needed for this particular type of takedown.

Think about this: If you were to sweep your opponent, would you sweep close to the foot, at the calf, or at the thigh? You would sweep as low as possible, close to the foot, because this would give you the longest lever arm (the distance between the applied force and the pivotal point; your opponent's leg in this case), allowing you to use less force.

The Catch-22

Mass can be described as the quantity of matter in an object. More mass = more powerful strikes. But mass is also a measurement of the inertia or "sluggishness" in starting, stopping, or changing the object's state of motion. More mass = more difficult to set in motion.

Newton's Second Law Of Motion states that *the acceleration of an object is directly proportional to the net force acting on the object, is in the direction of the net force, and is inversely proportional to the mass of the object*. In simple terms this says that, whereas force tends to accelerate things, mass tends to resist acceleration. So, a heavyweight must use more force than a lightweight to set himself in motion.

Newton's Third Law Of Motion states that *whenever one object exerts a force on a second object, the second object exerts an equal and opposite force on the first*. If you hit somebody, technically you will get hit back by his block with an equal amount of force. But if you will get struck back as hard as you strike your opponent, then what's the benefit of striking at all? And why is it that the person executing the strike seldom gets injured, while the person absorbing the strike does? The answer lies in your choice of target, and in the physical build of the striking weapon:

- First, you must strike a target that is likely to cause injury.

- Second, you must ensure that your striking weapon is structurally stronger than your target. That's why it is important to curl your toes back when throwing the front kick, to double up your fists tightly when throwing a strike, and to block a leg kick with the muscle slightly to the *outside* of your shin, and not with the bone.

To give your strikes penetrating force, the energy must be focused over a small surface area. Using your elbow rather than your hand to block a kick, for example, has two advantages:

- There will be more penetrating force, because of the smaller area of your elbow.

- You will stay protected better, because your hand can be kept in the high guard position.

Blocking a kick with the elbow.   **Blocking a kick with the hand.**

To sum it up, we can say that when striking and blocking, you should use parts of your body that are tolerant to impact, and strike parts of your opponent's body that are not.

The speed and power of spinning techniques

When we say that somebody is fast, do we mean that he can strike you before you can strike him, or do we mean that his strike travels fast on impact? The shortest distance between two points is a straight line, so a straight strike will always beat a looping strike. But this does not necessarily mean that a straight strike is faster on impact, or that it does more damage. Speed can be classified in two ways:

      1. The time your strike takes from start to finish.
      2. The actual speed in miles per hour.

A strike that is fast from start to finish has a strategical advantage, whereas a strike that is fast in miles per hour has a power advantage. This section will focus on speed in miles per hour, and why it makes striking powerful.

Distance is an interesting concept that is often deceiving. We normally think of distance by measuring the actual footage between ourselves and our opponent, or between our "weapon" and the target. Thus, distance can be decreased or increased by taking a step forward or back. If you have a lot of distance in which to accelerate your strikes, they will be more powerful on impact. This is because momentum = mass (weight) X velocity (speed). If you are very close to your opponent, your strike or kick cannot benefit fully from distance.

Keith throws a round house kick from three different ranges, impacting with the instep, shin, and knee. Aside from the fact that the knee and shin are powerful striking weapons, the longer distance of the round house kick with the instep enables Keith to build maximum speed on impact.

Because of our inherent physical characteristics, one might think that the maximum distance from which to land a technique successfully is limited to the length of our arms or legs. But if you understand the physics of rotational speed, you will find that distance can be increased far beyond your physical reach, and will therefore give you a tremendous power advantage. ***Rotational speed***, also called revolutions per minute, is how long it takes for a rotating strike to travel a complete circle. What makes this interesting is that although the rotational speed is the same everywhere in a rotating system, the ***linear speed*** is faster the farther from the center you get. If you place a record on the player, the slowest speed (in miles per hour) will be in the center and the fastest speed will be at the edge, even though the revolutions per minute remain the same. If you study traditional weapons, you probably have a *bo* staff at home. This simple staff has the advantage of a six foot reach. When performing your forms, you have probably learned how to rotate the bo staff above your head, either by holding it in one end, or by holding it in the center. The pivot point (where your hand is) will rotate very slowly in relation to the tips, yet the rotational speed stays the same throughout the rotating system.

Martial artists utilize many techniques that involve spinning. To maximize your speed, you should impact the target as close to the tip of the rotating system as possible. The spinning back fist and spinning heel kick are good examples. Yes, the shortest distance between two points is a straight line. But because spinning techniques don't follow straight lines, they also move through a distance far longer than the reach of your arm or leg, and therefore enable you to build a great deal of speed.

Rotational speed can be increased even more by jumping simultaneous to kicking. An airborne kick has less friction to deal with. In addition, if you contract your body throughout the spin until it is time to release the kick, you will speed up the spin even more. This is called ***conservation of angular momentum***. In simple terms it means that a compact body will spin faster than one that is not. Test this by grabbing a set of light handheld weights, 3-5 pounds (you can do this exercise without weights, but the increase in speed won't be as apparent). Hold your arms outstretched horizontally to your sides. Stand on one foot and set yourself spinning. Quickly pull your arms in toward the center of your body. Did your rotational speed increase automatically? We can now change the speed in a jumping and spinning technique just by varying the shape of our body.

The same principles of physics that apply to empty hand combat also apply to armed combat. Weapons are called "arms" because they extend our reach and aid combat efficiency. A stick, for example, will extend your reach by a few feet. In general, an armed attacker has the advantage over an unarmed opponent for several reasons. First, a weapon has no sensitivity; you don't risk injury to yourself the same way you would if striking with your fist. Second, a weapon enables you to reach your opponent even though he can't reach you. Third, because of the additional distance to the target, you can gain more speed and therefore more power. When striking with a stick, the highest speed is at the tip.

In both the horizontal strike and the downward strike, Martina impacts as close to the tip of the stick as possible. This maximizes her power, and gives her a safety advantage of maximum distance from her opponent.

There is an optimal length to a stick, and you must weigh the benefits in reach against the disadvantages of a weapon that is too long. A very long stick is difficult to maneuver because it has much of its mass away from the center. Although a single handed grip at one end enables you to build a high speed at the tip, holding the stick with both hands enables you to employ a push-pull motion (torque), so there is a trade-off depending on what you wish to achieve. The farther apart your hands are on the stick, the less force (muscular strength) you need in order to derive the same amount of torque (twisting leverage). Torque can be used when disarming an opponent with a stick. If he holds the stick at one end, and you grab the stick at the other end, technically your positions are equal and it becomes a matter of strength. However, if you grab the stick with both hands and start a push-pull (rotation), you can utilize the concept of torque to disarm a stronger adversary.

Martina uses the push-pull principle to take the stick from her opponent.

When blocking with the stick, is it be better to block with the middle or with the tip? Why? Although the tip is a better striking weapon, the middle is a better blocking weapon because it is closer to your hand. If you block with the tip, the force of your opponent's strike may create enough torque to twist the stick from your grip. **Torque = force X lever arm.** When the lever arm is short (from your hand to the middle of the stick, for example), for a given force the torque will be less than if the lever arm is long (from your hand to the tip of the stick). However, if the block serves an offensive purpose (as a strike to your opponent's knee when he is kicking, for example), more damage is done if you impact with the tip, as this is where the highest linear speed is.

When confronted with a weapon attack, we have a tendency to get hypnotized by the weapon. We tend to stay at long range in an attempt not to get struck. But keep in mind that your opponent can use distance to increase his power. If you understand physics, you may be able to neutralize his attack by moving to close range. Physics, in itself, is neither good nor bad. It can neither be given to you, nor taken away; it applies equally to all people at all times. It's how you use it that makes the difference.

Rotational inertia

If two fighters meet and attack each other head on, the stronger fighter will win. This is because the stronger fighter has the ability to back his opponent up and to dominate the space in which they are fighting. The intelligent fighter must therefore rely on circular movement both when fighting offensively and defensively. When you are the aggressor, using angled attacks instead of straight attacks will give you more open targets. Angled attacks are also likely to confuse your opponent, making it more difficult for him to defend against the attack. In turn, when you are the underdog and need to be on the defensive against an aggressive opponent, circular movement makes it more difficult for your opponent to land a good strike.

When fighting in a confined area, whether it is a ring in a tournament, a room in your house, or a confined area outdoors, the person who dominates the center will usually dominate the fight. The center is a strong position, because you have freedom of movement in every direction, forcing your opponent to expend energy by moving around you. If you fight empty handed, the center is stable against techniques designed to unbalance you. If you utilize a weapon, the highest impact power will be at the tip of the weapon. This enables you to fight from a distance, using very short and economical moves with the hand that is holding the weapon.

Because power relies on speed, you will want to strike your opponent with as much speed as possible. If you are too close to your opponent when throwing the spinning back fist, for example, it may result in striking him with your elbow instead of your fist. Aside from the fact that the elbow is a vicious striking weapon, it will not be able to generate the same speed as the fist, which is about twice as far from the axis of rotation when your arm is straight. This is also why it is important to extend your arm fully upon impact.

The ability to build speed throughout the spin also has to do with *rotational inertia*. Again, inertia means resistance to change. Rotational inertia means *resistance to change in an object that is rotating* (spinning back fist/kick, for example). How, then, do we overcome the rotational inertia of a spinning technique?

Rotational inertia depends on the distribution of mass with respect to the axis of rotation. This can be demonstrated with a hammer. If you balance the hammer on your finger tip, which is easiest, to balance the hammer on its head or on the shaft? Try it!

On the shaft is easier because the head-end is heavier. Most of the mass is therefore farther away from the center of rotation (your finger). The greater the distance between the bulk of an object's mass and its axis of rotation, the greater the rotational inertia. This is also why it is easier to balance on a beam or tightrope when you carry a pole. The longer the pole, the easier it is to balance, because the pole has much of its mass away from its rotational axis (center).

A spinning technique that has a great deal of rotational inertia is difficult to set in motion or to *accelerate*. A technique that can't accelerate will lack power (more acceleration = more speed = more power). We, as martial artists, must therefore find a way to overcome the rotational inertia when initiating a spinning technique. This can be done by keeping the bulk of your mass as close to your center as possible. Keeping your arms and legs tight to your body allows you to spin faster, and when it is time to release the technique, it will have more power through this faster speed. Anything that is sticking out from your body will resist the rotation. When initiating the spinning back fist, then, you must keep your arms tight to your body, and not extend your arm until it is almost time to land the strike. When resetting after the technique is complete, you should again contract your arm to the tight position. This will aid you in the spin back to your fighting stance. In addition to speeding up the technique, keeping your arms tight leaves fewer targets exposed.

Keith slips a jab and starts the rotation of his upper body. Note how his arms stay tight throughout the spin. Not until it is time to release the strike, does Keith extend his arm.

Whereas rotational inertia aids balance, it also counteracts speed. The potential loss of balance when throwing the spinning back fist, however, is not large enough to offset the benefits of the extra speed. In the pictures below, Martina Sprague can increase the power of her strike by sequentially extending her arm and using a high rotational speed. By extending her arm the moment before impact, she will impact with her fist, where the highest linear speed is. High speed = high power.

Martina impacts with her fist (the tip of the rotating system) where the highest linear speed is.

It is interesting to note that if you impact with the side of your fist, rather than the back of your hand, you can get an even greater power advantage. This has to do with the direction of the strike and with the resultant. When impacting with the side of the fist, the elbow comes up and points at the target, and the strike is then extended in the direction of the elbow, so there is no conflict with the direction of energy. When impacting with the back of the fist, however, the elbow is pointed slightly downward, so there is a slight conflict with the direction of energy. According to the principles of physics, connecting with the side of the fist is the more powerful of the two, providing that the striking weapon is structurally sound.

Unlike the spinning back fist or spinning heel kick, which are comprised of circular motion only, the spinning back kick is comprised of both circular and linear motion. When initiating this technique, you are utilizing circular motion to build momentum for power. But when landing the technique, your leg will be extended straight toward the target (linear motion). Unlike the spinning back fist or spinning heel kick, which strike their target from the side and in the direction of the spin, the spinning back kick strikes its target straight (like a jab or rear cross). When there is a lot of rotational inertia, it is difficult to change direction from circular to linear. By keeping your leg tight throughout the spin, this last moment change in direction right before impact will take less effort.

Keith starts the spin for the spinning back kick (circular motion), then extends his leg straight to the target (linear motion).

So, in addition to speeding up the spin, keeping your leg tight throughout the spin also speeds up your kick once it is time to extend it straight into the target.

Summary and review

There are two types of speed:

- The speed of the strike in miles per hour (actual speed).
- The speed of the strike from start to finish (how long it takes to land).

How fast you can build speed is also important. The force is equal to the mass times the acceleration.

$$\boxed{\textbf{Force = mass X acceleration}}$$

The lead weapon is faster than the rear in the sense that it reaches the target faster, but it is not necessarily faster in miles per hour. This is because it doesn't have the benefit of building speed through distance. Speed also depends on how comfortable the combination is. If your combination employs strikes in different directions, your strikes are not likely to be as fast as if all strikes are in the same direction.

Speed when striking

Speed is the second element of the momentum equation. Speed has both a power advantage and a strategic advantage. If the speed of the strike is significantly greater than the weight of the fighter, the strike is likely to result in a knockout or break rather than a knockdown. But speed without proper body mechanics does not give you maximum power. You can avoid starving the strikes of power by learning body mechanics first.

Linear speed:

1. Two strikes in different directions with the same hand (example: jab/lead hook) are slower than two strikes in the same direction with the same hand (example: double jab). This is because it takes energy to stop the movement and change direction.

2. Returning your hand to the point of origin when throwing double strikes in different directions, will help you change direction easier. This is because there is less rotational inertia when your hand is close to your body.

3. When throwing double strikes with the same hand, your body, and not just the arm, must reset between strikes. You can increase speed by resetting the body halfway only, but there are some drawbacks in power. If you are very close to your opponent, you may not have the opportunity to take a step for momentum.

4. The speed in combinations can be increased by using the pulley-effect, where one strike helps the other. You are now relying on both sides of your body, rather than just the striking side.

5. The tighter you keep your arms to your body, and the more you rely on body momentum, the faster and more powerful the strikes will be.

Circular speed:

1. Circular motion can make your strikes faster. This is because there is no start/stop required. But you must be careful not to paw; the strike must be absolutely straight on impact, or the resultant force will split in different directions.

2. Circular motion can be used between strikes with the same hand: between two uppercuts, for example. To avoid splitting the power, the circle should come at the initiation of the strike rather than on impact.

3. The circle should be elongated, or the strike will travel through a distance that is uneconomical.

4. Strikes that use large circles allow you to strike with both hands simultaneously without contradicting the principles of motion. This allows you to take many targets in a very short time.

5. When alternating between right and left upward strikes, a short bob and weave will help you speed up and set for the strikes. The bob and weave acts as a circle that enables you to build momentum.

6. The speed in shutos (inward to outward) can be increased by using a small circle when reversing direction. The same is true for hammer fist strikes.

A small circle after the hammer fist allows you to throw the back fist with speed.

7. A variation of the soft palm strike is the finger whip. This strike relies on speed, and therefore high kinetic energy. The finger whip can be used against the eyes or groin.

Backhand finger whip to the eyes.

Spinning strikes:

1. Spinning strikes rely on a complete or nearly complete circle, and therefore differ from looping strikes. Examples of spinning strikes are the spinning back fist and the spinning elbow.

2. The arms should be kept tight throughout the spin. This allows you to spin faster, because there is less rotational inertia. The rotation should be initiated in the body with extension of the arm just prior to impact.

3. The target should be impacted as close to the tip of the spinning weapon as possible. This is where the highest linear speed is (speed in miles per hour).

4. Momentum can be gained by taking a step simultaneous to spinning. The step can be just a few inches across the centerline, or it can be a bigger step. There should be no pause between stepping and initiating the spin.

Step across your centerline to accelerate the spin and place your body weight behind the blow.

5. Power can also be increased by increasing the rate of the spin, and by relaxing your arm and snapping the strike back from the elbow after impact.

6. Because spinning strikes connect from the side, they can be set up with any other looping or spinning strike in the same direction. This allows you to continue the momentum you have already started. The second strike will be faster than the first, as long as you take advantage of the momentum from the first strike.

7. If you miss with a spinning strike, take advantage of the momentum by launching another looping or spinning strike in the same direction.

If you miss with a hook, you can continue the motion into a spinning back fist.

Speed when kicking

By using the laws of physics, you can make your kicks faster without using more energy. Pay attention to correct mechanics. It is easy, but not at all necessary, to sacrifice good mechanics for speed. Momentum in spinning kicks (spinning back kick, spinning heel kick, tornado kick) is gained mostly through the spin. The farther you impact from the center of rotation, the higher the speed, and the greater the power. The exception is the spinning back kick, which relies both on a spinning and a straight motion.

Chambering and point of origin:

1. Bending the knee (chambering) will decrease the inertia of the kick. Less inertia means less effort, and therefore a faster kick.

2. The kicking foot should be brought back to the point of origin between each kick. Check that your stance is not too narrow or too wide. Incorrect stance makes it difficult to attain balance and accelerate successive kicks.

Speed in combinations:

1. Alternating kicks usually flow better than kicks thrown from the same side. This is because your body is allowed to reset between each kick. Better flow results in higher speed.

2. The same is true for kicks and strikes in combination. Thus, a lead leg kick followed by a rear hand strike will flow better than a lead leg kick followed by a lead hand strike.

3. Try to use the reverse movement (when resetting your body's balance) from one kick to launch the next. This enables you to decrease the beat between kicks.

Stepping in the direction of the kick:

1. Stepping in the direction of the kick adds the speed and weight of your body to the kick. The step can be accelerated by pushing off with your rear foot. Be careful not to jam the kick. Jamming will result in power loss, or in a push rather than a powerful kick.

A step does not have to be straight toward the target. In a leg kick, a lateral step will accelerate the kick and place your body weight behind it.

2. Eliminate unnecessary moves. Stepping with your lead foot and kicking with your lead leg requires a three-count (a step with your lead foot, a step with your rear foot to center your weight, and a kick with your lead leg). Stepping with your rear foot and kicking with your lead leg requires only a two-count (a step with your rear foot, which simultaneously centers your weight, and a kick with your lead leg), and is therefore faster through economy of motion.

Positioning for kick and distance:

1. Speed in kick combinations can be increased by planting your foot in position for the follow-up kick. For example, if you throw a rear leg front kick and follow with a lead leg round house kick, you can gain momentum and eliminate an extra step by planting your foot slightly to the side and in position for the round house kick.

2. Extra steps can be eliminated by using kicks that are the most appropriate for the current distance. If your front kick knocks your opponent back, following with a side kick may be better, faster, and more economical than following with a round house kick.

Speed in defense

Successful defense must be timed to your opponent's offense. If you react too slowly, you will get hit; if you react too quickly, you will also get hit. You can therefore not choose to speed up your defense the way you can choose to speed up your offense. Speed in defense should focus on launching the defensive move as late as possible (this allows you to keep a solid foundation and focus on offense longer), and to be as explosive as possible (this creates more power and makes up for the time lost through launching your defensive move late). You should also focus beyond the defensive move itself, so that you can speed up your counter-attack.

Body weight for speed:

1. Short defensive moves along your centerline (inward parries, for example, or short inward forearm blocks) allow you to use body weight to accelerate the move.

2. Don't reach for your opponent's strike; let it come to you. The closer your defensive weapon is to your center of gravity, the less inertia you have to overcome.

Circular moves for speed:

1. Circular moves (double parries, or bobbing and weaving) allow you to gain speed without stopping or resetting the technique.

2. The motion of a strike or kick can be utilized to launch a defensive move. For example, the motion of a hook can continue into a bob and weave to avoid your opponent's counter-strike. If your opponent doesn't throw a counter-strike, the motion of the defensive move can be used to build momentum for offense.

Speed in throws and takedowns

The primary principle of taking an adversary down is balance manipulation. Balance can be manipulated easiest by applying the concepts of rotational speed and torque. Rotational speed makes it difficult for your opponent to shift his center of gravity to retain balance. Once the takedown or throw has started, the circular motion will increase the momentum and keep your opponent's mind away from the escape or counter-attack. Torque allows a smaller person to take advantage of leverage to overpower a stronger adversary.

Circular motion:

1. Takedowns and throws work best when you apply circular motion. Circular motion is more difficult for your opponent to resist than linear motion. Circular motion also allows you to conserve energy by staying in the center of the circle. Your opponent must now fight against the forces that throw him off balance.

2. Yet, a properly executed throw is a combination between circular and linear motion: circular in the sense that you pivot your body, and linear in the sense that the focus of the throw is a diagonal line downward.

If a takedown is circular only, your opponent will move around you without going down (left). If it is linear only, you will end up carrying all his weight (right).

3. Many throws involve a step behind your opponent to unbalance him. Takedowns can generally be done without the step, with the unbalancing effect coming from a large circle with a high rotational speed. Keeping a wide stance will help you with this.

4. When applying the circular foot movement, only one foot (the sweeping foot) should move. Moving the foot that is the pivot point may result in inability to finish the circle.

5. The circular takedown employs two circles: the wide, sweeping circle of your foot, and the tight circle of your body. While your foot covers a large area, your body should be centered with the controlling technique held as close to your body as possible.

Don't allow the controlling technique (the joint lock, in this case) to separate from your body (left). Keeping it close will give you more control (right).

Torque:

1. In many hip and body throws, your positions are initially equal. You must therefore be the one who initiates the technique, preferably with some type of leverage.

2. When somebody grabs you, your first concern should be safety. You must therefore assume a superior position from the start. By manipulating your opponent's elbow, you will be able to turn his back toward you.

By using the leverage (torque) of the elbow, you can turn your opponent to an inferior position.

Speed when grappling

The same principles of circular motion and torque that apply when standing up, apply on the ground. Because of your lesser speed and flexibility when on your knees, you may need to rely on slightly different movement. Placing your opponent in an inferior position on his stomach should be the ultimate aim. Never jeopardize your own position by allowing your opponent to grab your arm, leg, or head. Much of grappling relies on leverage (torque). Because torque is the lever arm times the force, the longer the lever arm, the less force needs to be applied to move your opponent into the desired position. You should therefore look for lever arms that are as long as possible.

Circular motion:

1. When kneeling, the motion of the takedown will be somewhat restricted, because the ground will act as a barrier. Your knees, and not the balls of your feet, should now be used as pivot points.

Circular takedown from kneeling position.

Forward throw from kneeling position, as defense against a rear choke.

2. Being on the balls of your feet rather than your insteps, keeps you from getting pushed off balance backwards. It also helps you get to a standing position quickly.

On the insteps   **On the balls of feet**

Torque:

1. In most takedowns, your opponent will end up on his back or side. An opponent can be turned to his stomach by using torque against the natural movement of the joints.

Twisting joint lock to turn opponent to his stomach.

2. If you have hold of your opponent's leg (as when you have caught his leg in the crook of your arm and taken him down backwards), you can utilize torque to turn him over on his stomach by turning his foot toward his centerline.

Turn the opponent to his stomach by turning his foot toward his centerline. If you turn the foot away from the centerline, there will be too much weight for the ankle to handle, and you may cause severe damage to the ankle.

3. If you are on your back with your opponent kneeling between your legs, you can use the strength and leverage of your leg against his neck to unbalance him. You can then execute a breaking technique against the elbow.

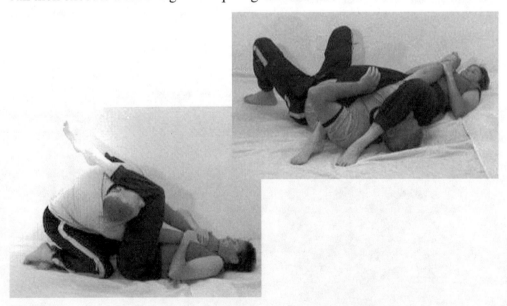

Use your stronger leg against your opponent's weaker neck. Once you have unbalanced your opponent, use torque to break his arm.

Friction and rotational speed quiz

1. How can a martial artist reduce friction without sacrificing balance?

Different substances have different amounts of friction. To the martial artist, friction is advantageous sometimes, but not always. For example, if engaged in a confrontation, you will be better off fighting from a high friction dry surface area than on ice. However, the less friction there is between the technique and the substance through which it moves, the easier it will be to gain speed. You can decrease friction by allowing your body to turn freely in the air. As long as there are no opposing movements in body mechanics, balance does not need to be sacrificed.

2. The shortest distance between two points is a straight line. How can you utilize the benefits of circular motion without increasing distance?

You can't! A circle will always cover a longer distance than a straight line. But is it really distance we are concerned with? Keep in mind that fighting is a give/take situation, where sometimes you must give something up in order to gain a bigger advantage. Circular motion enables you to increase your speed by eliminating the start/stop movement. The benefits of a sufficient increase in speed will offset the tiny bit of extra distance you need to cover. An elongated oval will enable you to increase your speed significantly without increasing distance that much.

3. Newton's Laws Of Motion often seem to contradict our efforts in gaining power. For example, you can't strike somebody without being struck back equally hard by the target you strike. How do you inflict damage on your opponent, yet remain unharmed by Newton's Third Law Of Motion?

When striking an opponent, you must use a part of your body that is structurally stronger than the target you are striking.

4. Why is it easier to do sit-ups when your arms are extended in front of your body than when they are clasped behind your head?

When your arms are extended in front of your body, your center of mass is closer to the rotational axis, thus less rotational inertia.

5. Why do we bend our legs more when running than when walking?

Bending your legs helps your body contract. This reduces the rotational inertia and allows you to speed up the run.

6. Why does the dachshund have a faster stride than its owner?

Rotational inertia is dependent on how far the bulk of the mass is from the axis (center of rotation). Short legs have less rotational inertia than long legs, and can be moved quicker. To demonstrate, grab a set of light handheld weights (3-5 pounds). Hold the weights in your hands with your arms extended straight out to your sides. Spin on one foot, then quickly pull your arms in tight to your body. Did the speed of the spin increase automatically? You may even have lost your balance as a result. So, you see, when your body contracts, it is easier to gain speed in a spinning technique. This is also important when performing jump kicks that require a turn of your body in the air (jump spinning back kick, jump round house kick). You can change your rotational speed by making variations in the shape of your body.

Glossary

Acceleration---Changes in speed and/or direction. An object that is in motion *and* changes its direction (a car driving up the clover leaf on-ramp to the highway), will accelerate even if there is no change in speed. You can feel that acceleration is taking place by the way your body lurches forward, back, or sideways.

Adding Momentums--When two objects move toward one another and collide, the force of impact will be stronger than if only one object is moving, providing that the speed is the same in both cases.

Conservation Of Angular Momentum--Whenever a rotating body contracts, its rotational speed increases. A rotating body will cover the same area within a specified time frame. A rotating body which is "spread out" will cover a larger area than one which is contracted. Therefore, the contracted body will need to spin faster in order to cover the same area.

Friction--Resistance of motion between two solid surfaces, liquids, or gases. Friction is less in air than on water or on ground.

Inertia---Resistance to change in motion. An object at rest tends to stay at rest; an object in motion tends to stay in motion.

Linear Speed--The tip of a rotating system will move faster than the center, even though the *rotational speed* (revolutions per minute) is the same throughout the system.

Mass--The quantity of matter in an object. When acted upon by gravity, we can use mass interchangeably with weight.

Momentum--The product of the mass of an object and its velocity. The heavier the object, and the faster it travels, the greater the momentum.

Newton's First Law Of Motion--An object in motion tends to stay in motion unless acted upon by some outside force. Martial artists should strive to throw *continuous* combinations, rather than many single strikes.

Newton's Second Law Of Motion--The acceleration of an object is directly proportional to the net force acting on the object, is in the direction of the net force, and is inversely proportional to the mass of the object. Whereas a big force produces a large acceleration, a big mass resists acceleration.

Newton's Third Law Of Motion--To every action there is an equal and opposite reaction. You cannot hit someone or something without being hit back with an equal and opposite force.

Resultant---The final sum of all vectors. When throwing a strike, we should strive to make the resultant as great as possible.

Rotational Inertia--Resistance to change in an object that is rotating. It takes a force to change the state or direction of rotation.

Rotational Speed--The number of rotations per unit of time (revolutions per minute). In a spinning technique, the revolutions per minute in the center is the same as the revolutions per minute near the tip. However, the tip will move with a faster *linear* speed, because it is farther from the axis of rotation. The rotational speed is the same regardless of how far you are from the center, but the linear speed is proportional to the distance from the axis.

Vector--An arrow symbolizing the strength and direction of a force. The longer the arrow, the stronger the force. For maximum power, all vectors must point in the direction of your strike.

Velocity--A measure of the speed of an object and its direction.

Impulse
Striking Through The Target

Have you ever witnessed a board or brick breaking competition where the practitioner was asked to break the bottom brick while leaving the others intact? Did he succeed? Did he use some kind of magical trick to do this? Did he cheat? It is in fact possible to strike the top brick and break the one on the bottom. Some people call this "extension of *ki*," or *internal energy*. I like to call it transfer of *ki*netic energy. Kinetic energy has the ability to penetrate the surface, which is one reason it is so damaging.

Your martial arts instructor has probably told you a number of times to strike "through" the target. Those who are less schooled in the martial arts may interpret this as *your fist is actually going into your opponent's body and through it, maybe even coming out the other side*. But in physics, this means allowing the *power* of the strike to extend beyond the physical surface and into the opponent's body.

Focusing a few inches beyond the physical surface allows Keith to extend the power of his strike through the target.

The definition of **kinetic energy** is *energy in motion*. This might have been demonstrated to you in physics class in school, where the teacher had a set of balls hanging on strings from a bar. All the balls are touching one another. The ball on the far end is then moved and released. When it hits the next ball, the ball at the other far end pops out. The balls in the middle, however, do not move. This is a demonstration of how energy can travel through a medium.

In martial arts, there is a fine line between snapping and *pushing*. If a strike is left in contact with the target too long, it will result in a push. But if a strike is snapped back too soon, it will lack proper transfer of kinetic energy; it will merely hit the surface without doing any real damage. In order to develop full power striking, you must first develop a feel for the *proper time of impact*. Working on relaxation will help you with this. A tense strike uses muscular power to reverse the direction of motion upon impact. A relaxed strike, on the other hand, allows you to sink the strike beyond the physical surface and through the target, and to use the target itself to aid in reversing the strike back to the guard position (point of origin).

It is my belief that if you want to develop power for full contact fighting or self-defense, you must train accordingly. Some touch sparring competitions penalize the practitioner for using too much contact. He should only mark his strikes with a light to moderate tap. When this is done repeatedly, we train our muscles to pull the punches. We judge distance wrongly, and do not extend the strike beyond the physical surface. Holding back interferes with proper energy transfer. You must therefore evaluate your situation and decide what your goals are. Do you practice for sports? Competition? The street? Kick-boxing? Reality fighting? If you do decide that you want to practice for sports/touch sparring tournaments, you can still develop power by practicing full power striking on bags or mitts a couple of times a week.

You must now find the fine line in between the snap and the push, where the striking weapon is left in contact with the target long enough to allow for proper energy transfer, but not so long that it results in a push.

Decreasing the time of impact in the grappling arts

We have already talked about how the **momentum** of an object will change if either the **mass** or **velocity** (speed) changes. We have also talked about how mass, speed, and momentum all relate to power. **Impulse** is defined as the *force times the time interval*, or as the *change in momentum*. A large change in momentum in a long time requires a small **force**. A large change in momentum in a short time requires a large force. For powerful striking, we should strive to produce a force that is as large as possible. When throwing a punch that suddenly stops or reverses direction upon impact, the shorter the time in which the momentum is changed (reduced to zero), the larger the force.

Let me ask you this: If you were to jump from a five story building, where would you rather land, on the concrete sidewalk below, or in a net that the fire department had brought out to your rescue? We know that it would hurt less to land in the net, but why? The reason is that the time during which your momentum changes (during which you come to a stop) is much longer when landing in the net than when hitting the concrete sidewalk. The net has "give," which slows you down little by little. When the time is long, the force is small, and you are more likely to walk away uninjured.

<u>Try this exercise:</u> Get up on a chair and jump off, landing on your feet. Why did you bend your knees? Because it increases the time of impact, and your body will gradually absorb the force. If you lock your knees instead, you will get hurt, even when jumping from a very low height. Don't try it!

Many martial arts employ throws. It is therefore crucial that the practitioner learns how to fall without getting injured. Aikido and Judo practitioners get thrown to the floor repeatedly, yet keep from getting hurt, because they have learned how to spread the impact over as large an area of their body as possible. The forward roll, for example, utilizes the hand, forearm, shoulder, and back to gradually and sequentially absorb the shock. This spreads the impact over an extended period of time. More time means less force. It is therefore possible to fall from a considerable height, or to be thrown at a considerable speed, without getting hurt.

In the forward roll, the force is absorbed sequentially over a relatively long time.

In addition to helping you absorb the impact of a fall, a rolling technique also gives you considerable momentum, which can be used to gain distance from your opponent and get back on your feet.

The second way to absorb the shock of a fall, is to absorb the force all at once by spreading it over as large an area of your body as possible.

Note: The principle of spreading out to absorb the shock applies only to situations that involve a solid surface, as would be the case in most fighting situations. If falling into water, the opposite is true. The force will be reduced by allowing your body to penetrate the surface. Consider a belly splash off a fifteen foot high springboard. Would it hurt? If you made yourself smaller instead, allowing your feet or head to penetrate first, you would come through the fall unharmed.

The amount of control you have over the situation will determine which type of fall you should use. In general, it can be said that if the fall is a result of your own initiative, as would be the case when a fighter is escaping a controlling technique by executing a forward roll, the principle of absorbing the shock gradually and sequentially should be used. The force is absorbed over a long period of time and is therefore lessened **(concept: Impulse = Ft)**.

If the fall is a result of your opponent's initiative, as would be the case when a fighter is picked up and thrown, he may not have the option of a rolling technique to break the fall. He should then rely on the second principle of falling; that of spreading the impact over as large an area of his body as possible **(concept: pounds per square inch)**. The practitioner of this type of fall must keep all vital areas from touching the ground.

When utilizing the backward break fall, your chin must be tucked down toward your chest to keep your head from hitting first. You should also make sure that your arms are extended and not bent, as you would otherwise absorb the force on your elbows. Slapping the mat simultaneously or slightly before landing, and at an angle away from your body, will help you absorb the force over a greater surface area.

Keith executes a backward breakfall with a slap at a forty-five degree angle away from his body.

When utilizing the side break fall, the force should also be absorbed over as large an area as possible. Note, in the pictures below, how Keith lands on the whole side of his leg and body, and how his arm slaps the mat at an angle away from his body.

The next fall is to the front from a standing position. When utilizing this technique, we have a natural tendency to want to catch ourselves on our hands and knees. In order to absorb the shock of this type of fall, the hands and forearms must slap the mat a fraction of a second before the body lands. If you land on your wrists, elbows, or knees alone, the force will be absorbed over a very small surface area which is also structurally weak. This will produce a greater force per square inch, with a greater risk of injury.

Note, in the picture at right, how Keith prepares to absorb the shock along the palms of his hands and forearms. If the fall was executed from a standing position, Keith's knees would not be touching the ground, and the weight of his body would be supported by the toes and forearms alone.

Decreasing the time of impact in the striking arts

In a stand-up fight, going *with* the motion of your opponent's strike will increase the time of impact, thus reducing the force. Moving into your opponent's punch will hurt more than moving away from the punch. When working on offense, you will therefore want to time your strikes to your opponent's forward motion, making him walk into your strikes. This will change the momentum in a short period of time, thus increasing the force of the blow. For best results, the force should be delivered in as short a time as possible.

> **Change in momentum = force X time**
> **or**
> **Impulse = Ft**

Impulse = F**t**   Impulse = **F**t

In the picture on the left, Keith executes a rearward slip to minimize the force in his opponent's blow. In the picture on the right, he makes the mistake of moving forward and into his opponent's strike.

The force can be further increased by striking a target that has very little give. In the picture below, the target is immobilized by pulling on the opponent's arm, while simultaneously kicking to the body. Because the body is kept from moving back, the time of impact is shortened, and power is increased.

Martina increases the power in her kick by pulling on her opponent's arm.

The power in a takedown can be increased by applying pressure against the natural movement of the joints. This will immobilize your opponent, giving the target area very little give. If you throw a strike now, power is increased because of the short time of impact against a target that can't move with the force.

A shoulder lock immobilizes the target. A knee strike to the head would now be devastating.

Since the joints are inherently weak, a smaller person can use a joint lock to defeat a bigger adversary. By immobilizing the joint and relying on the concept of *impulse*, he can increase the effectiveness of the technique.

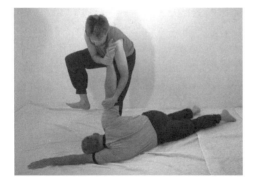

Martina uses the ground in conjunction with a shoulder lock to immobilize her opponent. A stomp to the head would now be devastating. When the head has no place to go, the power is absorbed in a very short period of time.

198   Fighting Science

Impulse can also be used to break free from a lock-up. A chain will break at its weakest link. In the pictures below, the opponent's arms around Martina's body can be viewed as a link in a chain. The weakness is where the hands come together. By forcefully spreading her arms, Martina is able to break the grip, and continue with a counter-attack. In order for a smaller person to benefit from this technique, the moves must be executed dynamically. Especially in grappling, where you are in actual physical contact with your opponent, any small move or "intention" can often be felt at its earliest stage. If a person fidgets or applies pressure gradually, it will be easy for the opponent to pick up on the move and increase the intensity of his attack. Breaking free should happen explosively and without any prior warning. The shorter the time, the greater the force. If the break simulates a push, you are not likely to succeed.

Martina escapes a rear bear hug by forcefully spreading her arms.

Reversing direction by "bouncing"

Karate practitioners who are into board breaking will be the most successful if they can reduce the momentum of their strike to zero in the shortest time possible. Stopping the momentum suddenly (as when your hand or foot impacts the board) produces a large force. An interesting concept is that if your hand or foot is made to "bounce" upon impact, the force is even greater. This is because the impulse required to bring your strike to a stop, and then reverse the motion by "throwing it back again" is greater than the impulse required to simply bring it to a stop. When your strike stops, the momentum changes to zero; when you reverse its direction, the momentum must again change from zero to a higher number, so the total change in momentum is greater than if you simply brought the strike to a stop. The force is increased because the time of contact is reduced. The impulse is also increased because the direction of the strike is reversed. The impulse, when reversing direction, is supplied by the opposing surface (the board), as your hand or foot bounces off it. Be careful not to use muscular effort to snap the strike back.

The momentum goes from 10 to 0.

Keith initiates a punch with a momentum of ten (for simplicity). At impact, the momentum changes to zero. The total change in momentum is therefore *ten*.

The momentum goes from 0 to 10.

When Keith retrieves his hand back to the guard position, the momentum goes from zero to ten, but in the *opposite* direction. The total change in momentum when throwing a punch that reverses direction upon impact is therefore *twenty*, or twice that of a punch that stops and does not reverse direction.

Kenpo karate practitioners rely on bouncing their strikes off of both their own and their opponent's body. Kenpo karate is often negatively referred to as a "slap art," where the slaps are merely looked upon as some kind of fancy show-off. Many Kenpo practitioners also claim that these slaps are in reality "checks" against the opponent's blows. But those who understand the underlying principles of physics, will be able to increase their power by bouncing their strikes off the target. This can almost be thought of as adding momentums. When you strike something, it strikes you back, thus producing the bouncing effect. Time of impact is lessened and power is increased.

When the Kenpo practitioner allows his hand to bounce off his own body, the strike is reversed by relying on the opposing surface (his body) and not on any muscular effort. This helps the practitioner build speed without getting arm weary. More speed results in an increase in power.

When the Kenpo practitioner relies on the target (his *opponent's* body) to provide the reversal of motion, the impulse, and therefore the power, is increased. If you subconsciously use your muscles to bring your hand to a stop and reverse its direction, the desired results cannot be attained. The impulse is created by bouncing your hand against the target. The snap should come from the *opposing* surface and not from within. The impulse must be exerted on the object by something *external* to the object. Internal forces don't count. This can be likened to a person sitting in an automobile and pushing against the dashboard. It doesn't matter how hard you push, you will have no effect on the speed of the automobile. This is because these forces are *internal* and act and react within the automobile itself. If no external force is present, then no change in speed is possible. A relaxed strike will hurt more than a strike that is tense, because a relaxed strike is allowed to penetrate the target (with minimum time of contact), and use the *target* to aid in the reversal of direction (an outside force), and not its own muscles (inside force).

The same principle can be used to reverse direction and increase power in spinning techniques. In the pictures at right, upon landing the spinning back fist, Martina's body is like a spring unwinding into a second spinning back fist with the other hand. Two things are occurring here: First, she uses the necessity to reset into a fighting stance to launch a second attack. Second, she gains power in the spinning back fist by reversing direction upon impact.

There is a difference between throwing two single strikes and throwing a two-strike *combination*. A combination utilizes the movement of the first strike to start the movement of the second strike. In other words, the strikes are helping one another. This is also a way to conserve energy. The shorter the movement, or the less time between strikes, the more energy is conserved, helping you to last longer during heavy physical exertion. As mentioned earlier, because of less *inertia* (resistance to change), it is easier to keep a combination going than to start and stop many single strikes.

Martina throws a double spinning back fist. Note how she utilizes the *back* of her hand, with her elbow angled slightly down. This strike would not be quite as powerful as the spinning back fist that utilizes the *side* of the hand (see section on "Rotational inertia" in Chapter 5).

Impulse when kicking

The principles of physics are not technique specific, and apply equally to kicking and punching. The moment of impact must be as short as possible when kicking, or the kick will result in a push, with the power dispersed into the target over a longer period of time. After the kick has landed, a quick reversal of the momentum will result in an increase in power. In order to make the moment of impact as brief as possible, you must retrieve the kicking leg as fast as possible, using the opposing surface to aid in the reversal of motion. This is accomplished by letting your foot bounce off the target.

When kicking, rely on the following principles:

- Raise your leg in the cocked position. If the leg is straight, you will have difficulty accelerating the kick. This is especially true if the kick employs circular motion (round house kick, spinning back kick), because the ***rotational inertia*** will counter-act your efforts to accelerate the kick. Other benefits of raising your knee high are easier rotation of the hips, and better protection against your opponent's counter-strikes.

- When extending your leg to kick, think of impact as coming slightly before your leg is fully extended. This allows you to extend the kick through the target, with the energy of the kick extending beyond the surface.

- Snap your leg back to the cocked position. The snap should come mainly from your opponent's body and not from the muscular control in your kick, thus creating a greater impulse by reversing the momentum. Make sure that your leg is snapped back in the same direction it originated from, or the greatest change in momentum cannot take place.

Note: Raising your leg (cocking it) and throwing the kick may seem like two separate moves, but they need to be synchronized into one fluid motion with no stop and restart of momentum.

Three types of impact

Successful fighting relies on the martial artist's ability to throw a powerful strike, and also on his ability to develop, understand, and use sound strategy. Every move should have a clear purpose of either set-up, knockout or breaking, or positioning and balance. We can generally say that there are three types of impact:

1. Stinging impact.
2. Shattering impact.
3. Pushing impact.

Let's look at *stinging impact* first. Depending on what you wish to achieve, it is not always wise to produce a blow at its maximum force. If you compete in point sparring, you will be disqualified for using excessive force. It may therefore be better to develop a strategy that will earn you as many points as possible. This is done through a series of set-up moves. When you have created an opening in your opponent's defense, you can strike with a quick but non-damaging blow. The shorter the time the strike is in contact with the target, the greater the force. But in point sparring, the force will be low regardless of time, because the fighter uses his muscles (inside force) to hold back the power and reverse the direction of the strike (see section on "Reversing direction by bouncing" in this chapter). This type of strike is likely to sting, but unlikely to do any significant damage. In a full-contact match, a strike with stinging impact is useful as a strategic move to irritate your opponent, create openings, and split his focus prior to throwing a powerful knockout punch.

Shattering impact is used to shock the body, as when breaking bones or delivering a knockout strike. Shattering impact is also used in board and brick breaking demonstrations. In this type of impact, the striking weapon should be in contact with the target a very short period of time, yet the practitioner must use extreme caution not to rely on his muscles (inside force) to snap the strike back after impact. There is a fine line here: in both stinging and shattering impact, the strike is "snappy" (quickly reversed), but shattering impact has more power. Why? Because the strike is thrown relaxed and relies on the target to aid in the reversal of motion. This type of strike does not take a lot of muscular effort, yet it is very powerful and damaging. The practitioner must understand that powerful strikes are snappy with a short duration of contact, but not so snappy that the strike is starved of power. Using muscular control to stop the strike and reverse its direction interferes with proper target penetration. A strike with shattering impact should rely on the target (outside force) to stop the strike and then, in effect, throw it back again (bouncing). The force is increased because the *impulse* (change in momentum) is doubled when the strike reverses direction on impact.

Bouncing of a strike is not always obvious. Let's say, that you want to knock your opponent out with a punch to the jaw. But because the head is lighter than the body, and the neck is an inherently weak part of our anatomy, the strike is likely to continue through the target, rather than reversing direction on impact. The head will move, but the body will be left behind. This is because of the inertia of the much heavier body. So, now you have both momentum, which creates wallop, and kinetic energy, which does damage. If you wanted to break a rib, you can again use shattering impact. The shorter the time of contact, and the faster you reverse the direction of the strike (change the momentum), the more damaging the strike will be.

Note: You don't *have* to have a bounce (*only* a short time of contact) with shattering impact, but if you do master the reversal of direction through bouncing, the force will increase.

Pushing impact is useful when you wish to reposition your opponent for a follow-up attack. This type of impact keeps your striking weapon in contact with the target longer, and allows you to move (push) your opponent back against a wall or the ropes of the ring. Pushing impact works best if your opponent (the target) has somewhere to go. If your opponent is already with his back to a wall, pushing impact is difficult to use, as well as strategically unsound. It would then be better to use shattering impact. If the target has no place to go, the time it takes for the strike to come to a stop is automatically reduced.

A push is not always obvious, as there are different degrees of pushing. A push does not have to take a long time; only longer than a strike thrown with shattering impact. Pushing impact is great for stealing balance, especially when your opponent's base is narrow (as when he is in the process of kicking). You can achieve pushing impact with a punch, but a strong kick, like a side kick or front thrust kick, is better. Pushing impact can also be used in a takedown/grappling situation, as when tackling an opponent preparatory to a takedown.

When using pushing impact, focus on carrying the *momentum* (not the kinetic energy) through the target. For example, a stepping side kick from long range allows you to build momentum, and to keep that momentum going until you have achieved the result of moving your opponent back. Depending on how dynamic your attack is, and because the human body is only capable of absorbing a certain amount of force, a technique using pushing impact may produce damage in addition to moving the opponent back. But this is outside of the principles of physics, and has more to do with human anatomy.

To sum it up, it can be said that:

- The force can be increased by decreasing the time of contact, or by increasing the change in momentum (the impulse).

- Stinging impact is not really related to the physics of power, although it does have a valid strategic purpose.

- Shattering impact has a lot of kinetic energy that goes through the target. Shattering impact is therefore the "deepest" of the three types of impact, and the most damaging.

- Pushing impact has a lot of momentum, and is useful both as a strategic move, as when unbalancing an opponent, and as a power move, as when keeping an opponent at long range with a kick.

- There is a fine line between the three types of impact. Unless you also practice, develop, and apply sound strategy, simply understanding the physics will not make you a master martial artist.

Impulse exercise

Definition: *Impulse is the change in momentum*, which is also the product of the **force** acting on an object and the **time** during which it acts. We will give the **force** the letter **F**, and the **time** the letter **t**. Therefore, **Impulse = Ft**. Or we can say that the **change in momentum = force X time**.

In order to **change the momentum** of an object, both the **force** and the **time** during which the **force** acts on the object are important. The **force** required to stop a punch will be smaller if it takes a long **time** to stop the punch, than if the punch is stopped quickly. Which hurts more, to hit a focus mitt that is soft, or to hit a granite slab? Slamming your knuckles into a granite slab is obviously going to hurt more. Why? Because the granite doesn't have give, and the amount of **time** required to stop the punch is very small in comparison to hitting the focus mitt, which has give and allows your hand to sink into it gradually. When the **time** required to stop a punch is small, the **force** will be large.

If you want to **change the momentum** of an object moving at a certain speed and coming to a complete stop (as is the case when throwing a punch), a *long* **time** is compensated for by a *small* **force**, and a *short* **time** is compensated for by a *large* **force**.

When we say that one thing is equal to another, we call that an *equation*. Let's look at the equation **change in momentum = force X time**. For simplicity, I will choose some numbers that are easy to work with. If I give the **force** the number **5** and the **time** the number **2**, then what would be the **change in momentum**? The equation would read **change in momentum = 5 X 2**, so the **change in momentum** would be **10**, right? Why? Because **5 X 2 = 10**.

Now, let's say that you are throwing a punch with a momentum of **10**. In order to stop that punch, we would have to figure out the difference between **10** and **0** (which is when the punch comes to rest). The difference between **10** and **0** is **10**, right? So, we could say that the **change in momentum** is **10**. In the previous example, I chose the number **2** for the **time**. If I were to decrease the **time** to **1**, then how much **force** would it take to stop the punch? The equation says that the **change in momentum = force X time**, or **10 = something X 1**. That "something" is the **force**. What number would you have to time **1** with to get **10**? Answer: **10**, of course! So, by comparing these two examples, we can see that by cutting the **time** it takes to stop the punch in half, we also double the **force**. The **time** went from a **2** to a **1**, and the **force** went from a **5** to a **10**. Are you with me so far?

So, the **force** of a punch will be greater when we decrease the **time** of impact. Another way to increase the **force** is to **reverse the momentum**. Let's say that the original **change in momentum** is **10**, because the momentum went from a **10** to a **0**, as in the example above. Now, let's say that the momentum goes from **10** to **0** and then back to **10** again, as would be the case if you threw a punch, let it come to a complete stop as it hit the target, and then let the target throw it back again in the same direction it came from (snapping the punch). Now, the **change in momentum** would be **20** instead of **10**, since **10** to **0** and back to **10** is equal to **20**. **20** is twice as large as **10**, right? Let's say that the **time** of impact is still **1**, then how large would be the **force**? Let's plug these numbers into the equation and see what happens.

Change in momentum = force X time. **20 = something X 1**. The "something" is your **force**. The "something" must be equal to **20** in order to balance the equation. So, we can see that by snapping the punch back in the same direction it came from and with the same speed, we can double the **force** (we went from **10** to **20**) and double the power in the punch. Simple, eh?

Summary and review

> **Impulse = force X time**

Impulse is the product of the force and the time. It can also be said that impulse is how much you change the momentum. Maximum change in momentum occurs when a strike bounces on impact. Whether this is beneficial or not depends on what you are trying to achieve in any particular situation, and on whether you and your opponent are on the offensive or defensive.

Impulse when striking

A push gives you momentum, but a snap gives you power. The shorter the time of impact, the greater the power. Whether you push or snap depends on your objective. Do you want to push your opponent back to create distance, or do you want to knock him out?

Short time=large force:

1. Force can be increased by shortening the time of impact, either by snapping the punch, or by pulling your opponent into the strike. Care should be taken not to use muscular effort to slow the punch and reverse its direction.

2. Whenever you add momentums, as when your opponent walks into your strike, the time of impact is shortened and power is increased. This is because the target does not have much "give" to move with the strike.

3. A strike that bounces off the opposing surface has a very short time of impact and therefore a great force. This can also be thought of as allowing the target to help you reverse direction.

4. The strike must be snapped back in the same direction it came from to avoid sliding against the target and splitting the resultant force.

If you drop your hand after landing a punch (middle), you risk splitting the resultant force, and maximum change in momentum cannot take place. Your hand should be brought back in the same direction it came from.

5. Try a double spinning back fist. Upon landing the first strike, allow it to bounce off the target and help you reverse momentum into a second spinning back fist with your other hand. The strike should be snapped back from the elbow. If the arm is kept straight, there will be more rotational inertia that will slow the strike down.

6. The power of the spinning back fist can be further increased by taking a lateral step across your centerline with your lead foot. This gives you the flexibility for maximum target penetration. Power can also be increased by increasing the rate of the spin.

7. A strike against an opponent on the ground has a short time of impact and a great force. This is because the ground has no "give," and doesn't allow the target to go with the motion of the strike.

Dropping your knee onto your opponent's head when it is in contact with the ground is a powerful technique with a high impulse. In addition, there is a great force per square inch, because of the small area of your knee.

Impulse when kicking

Impulse when kicking relies on the same principle as when striking. The shorter the time of impact or the quicker you reverse the direction of the kick, the greater the force. Timing the kick to your opponent's forward motion, so that he walks into the kick, also increases the force by shortening the time.

Short vs. long time of impact:

1. Kicking when your opponent moves into the kick allows the momentum to come to a stop in a short time, resulting in an increase in force.

Kicking an opponent who is against a wall will increase the force by shortening the time.

Grabbing your opponent and pulling him into the knee will increase the force by shortening the time.

2. Moving with your opponent's kick allows the momentum to come to a stop in a long time, and is therefore beneficial when taking a kick.

Going with the motion of a kick lessens the force by lengthening the time. Note how Cindy places her weight on her rear foot to make the kick slightly out of reach.

Reversing direction:

1. Reversing the direction of a kick after impact will double the change in momentum. This is one reason why you should snap your leg back to the chambered position after kicking.

2. To avoid splitting the resultant force, the leg should be snapped back in the same direction it came from.

Impulse in defense

Impulse in defense can be used to increase the force of your block, or to decrease the force of your opponent's strike. Many blocks serve an offensive purpose simultaneous to defense. You should therefore strive to make your blocks as snappy as possible. As long as you don't snap back prematurely, a block that is snappy gives you a greater force on impact than a block that "pushes" your opponent's strike to a stop.

Two flies with one swat:

1. If you can go with the motion of your opponent's strike, you will lessen the impulse and therefore the power. Try a rearward slip as defense against a strike. When your body comes forward again, use this to launch a counter-strike, relying on the weight and movement of your body for power. You have now accomplished defense and offense within the same technique.

2. When your opponent's strike misses, throw a counter-strike, ideally while he still has forward momentum. You have now "caught two flies with one swat"; you have decreased the force by increasing the time when going with the motion of your opponent's attack, and you have increased the force by decreasing the time when counter-attacking against his forward momentum.

Going with the motion of a leg sweep increases the time and lessens the force, and helps you launch a counter-attack.

Impulse in throws and takedowns

Whether you want to increase or decrease the impulse depends on whether you are the one throwing or the one getting thrown. A long time of contact means that the force will be absorbed sequentially by the body, and will therefore benefit the person being thrown. The thrower can try to counter-act this and increase the impulse by making the throw as violent as possible. In a training situation, mats will help decrease the impulse; in a street situation, a hard surface will increase the impulse.

Impulse when getting thrown:

1. If the throw or takedown involves circular or lateral motion, you can absorb the force by using the momentum to do a roll forward or on your side.

2. Some escapes from joint locks allow you to build momentum for a roll. An example is the escape from the twist lock. When your opponent applies the lock, tucking and rolling will decrease the pressure on your wrist and allow you to escape.

A forward roll will help you escape from the twist lock. By absorbing the force sequentially over your arm and shoulder, you will lessen the impact, even though there may be quite a bit of momentum.

3. Throws are generally more violent than takedowns, and may not allow you to absorb the shock gradually. Spreading your weight as wide as possible will enable you to use the principle of force per square inch to your advantage.

Impulse when throwing:

1. When you have the upper hand in a throw or takedown, you want the impulse to act on your opponent. The throw should therefore be as dynamic as possible.

2. Try to drop your opponent vertically straight toward the ground. This makes it more difficult for him to absorb the shock sequentially.

3. Most people have a natural tendency to want to catch themselves on their hands when thrown, which will work to your opponent's disadvantage. The force will now be absorbed over a small surface area, and will be very damaging.

Impulse when grappling

Because of both fighters' close proximity to the ground in a grappling situation, it is more difficult to do damage with a throw or repositioning technique on the ground than when standing up. Impulse, on the ground, can generally be increased through torque (leverage).

Impulse in joint locks:

1. An opponent can be repositioned through the use of a joint lock. A joint lock is most effective when done explosively. This gives you the advantage of the moment of surprise, the pain from going against the natural movement of the joint, and the impulse by not allowing your opponent time to go with the motion.

An opponent who is straddling you can be repositioned by using an explosive joint lock against the wrist.

2. If you fail to use explosiveness, the time of contact will be longer, allowing your opponent to overcome the inertia of his body and move with the lock.

If an arm bar is not applied forcefully, your opponent may roll on his side and grab his own arm in an attempt to lessen the force and break free.

3. When engaged in grappling, be aware of that joint locks are common, and that if you can go with the motion of the lock, you will have a lesser risk of sustaining injury.

Striking and grappling:

1. Strikes can be used in conjunction with grappling, and have the potential to do great damage. The best time to strike is when your opponent is in contact with the ground. This increases the impulse by limiting his ability to move with the motion of the strike.

2. A ground situation is in many ways similar to a stand-up fight, and you can use many of the same moves and principles. Pulling your opponent into a strike, for example, adds his momentum to yours and increases the power.

Grabbing your opponent around the neck and pulling him into a knee strike increases the power through a shorter time of contact.

Impulse quiz

1. What is meant by striking "through" the target, and how is it accomplished?

Striking through the target means allowing the *power* of the strike to extend beyond the physical surface. This is accomplished by relaxing to allow full extension of the strike, and by avoiding to subconsciously stopping the strike short.

2. What is snap? In what way is it beneficial? What is the danger of snapping back too soon?

A strike that is snappy utilizes minimum time of contact with the target. The benefit of a snappy strike is that the force is increased through a decrease in time. The opposite of snapping is pushing. When you push, you allow your hand or foot to stay in contact with the target longer, and power is reduced because the time is increased. We should strive to throw our strikes with snap, but there is a danger associated with snapping back too soon. A strike that is too snappy generally utilizes muscular control to reverse its direction, and therefore interferes with proper target penetration.

3. Why is power reduced when the time it takes to bring something to a stop is increased?

Impulse = change in momentum, which is also equal to force X time. When the momentum changes in a long period of time, the force (power) must decrease in order to balance the equation.

4. How can a martial artist taking a fall keep from getting hurt?

Many martial arts employ throws. It is therefore important to learn how to fall properly without risking injury. When the shock is absorbed gradually over a long period of time, the force is reduced and injury is less likely. This is accomplished by *sequentially* touching as many parts of your body to the floor as possible. A common example is the forward roll, where the impact is absorbed through your hand, forearm, and shoulder. Your head, which is especially sensitive to injury, never touches the ground.

5. Why is the impulse increased in a strike that is made to bounce upon impact?

Impulse = change in momentum. When a strike comes to a stop upon impact, the momentum is changed to zero in a specified period of time. When a strike reverses direction, the momentum is again changed. The *total* change in momentum is therefore greater than if the strike was simply brought to a stop. Bouncing aids in the reversal of direction, and in the total change in momentum.

6. To enhance power, bouncing or snap must come from the target (externally), and not from the practitioner's muscular control (internally). Why?

Using muscular control to reverse direction would be equivalent to sitting in a car and pushing against the dashboard in an attempt to move the car forward. We can also think of this in terms of Newton's Third Law Of Motion (to every action there is an equal and opposite reaction). When you hit something, it hits you back. A strike should be thrown relaxed and with full extension (penetration), and allow the *target* to throw the strike back again.

Glossary

Equation--A term frequently used in math and physics to indicate that one thing is *equal* to another.

Force--Any influence that can cause an object to be accelerated.

Impulse--Change in momentum, or the force times the time interval. The shorter the time, the greater the force, and vice versa.

Inertia--Resistance to change in motion. An object at rest tends to stay at rest; an object in motion tends to stay in motion.

Ki (or chi)--An expression often used in the martial arts to describe internal energy resulting from full synchronization between a person's physical and mental focus.

Kinetic Energy--Energy in motion (as opposed to potential energy). Kinetic energy is equal to *one half the mass times the velocity squared*. In Chapter 8 we will learn, in more detail, how this affects power.

Mass--The quantity of matter in an object. When acted upon by gravity, we can use mass interchangeably with weight.

Momentum--The product of the mass of an object and its velocity. The heavier the object, and the faster it travels, the greater the momentum.

Pounds Per Square Inch--The narrower the base of an object, the more pounds per square inch it will exert upon the opposing surface. A narrow base is less stable than a wide base, but it also has more penetrating capability. When falling, spreading your weight over as large an area as possible decreases the force.

Pushing Impact--This type of impact is used when attempting to move your opponent through a distance. Pushing impact may or may not have a high velocity. The time of actual contact with the target is relatively long with no attempt being made to reverse the momentum.

Rotational Inertia--Resistance to change in an object that is rotating. It takes a force to change the state or direction of rotation.

Shattering Impact--This type of impact is used when delivering a knockout strike, or when breaking bones, boards, or bricks. Shattering impact has a high velocity with a short time of actual contact with the target, and a quick reversal of momentum.

Stinging Impact--This type of impact is used when point sparring, or as a distraction or annoyance in full-contact fighting. Like shattering impact, stinging impact has a high velocity with a short time of contact with the target, and a quick reversal of momentum. However, because it relies on the internal muscular force of the fighter, it is not nearly as damaging as shattering impact.

Velocity--A measure of the speed of an object and its direction.

Conservation Of Energy
The Workload

I like to fight when I'm really tired. Why? Because if I can fight well when tired, I know that I can pull that little extra out of me when needed, and still survive.

It has been said that with sufficient stamina, you can wear down the most powerful adversary. Those of us who have competed, especially in a full-contact match, know to the fullest extent what it means to get tired. When you get tired, your years of training go out the door. Your techniques become worthless. Your power is reduced to zero. A blow, which you would ordinarily not have thought twice about, makes you stagger across the ring close to a knockout. Your legs are shaky, you struggle to maintain balance, and your will to fight is reduced to the point that you might even take a technical knockout willingly just to get out of it.

Through my own experience, I'd like to say that the single most important factor in winning a fight is to keep from getting tired. Even now, years later, I remember the frustration in my first kick-boxing match. When I got back to my corner after the first round, I was so tired I didn't think it would be possible to get back out for the second round, let alone to *fight!* Close to the end of the third round, my opponent was so tired that she was up against the ropes, back on her heels, both hands down by her waist. And the look on her face is still, all these years later, etched in my memory. I could hear my instructor yell from my corner: *"Your right, throw your right!"* It was a free shot; she was way open. *Would* have been a free shot, if only I hadn't been *so* tired. There was no way I could have mustered the strength to throw that right hand toward a possible knockout.

In a later fight, I was badly overpowered and constantly knocked back by a very aggressive opponent. I chose to cover up, to keep my cool. When I got back to my corner after the first round, they told me, "You can't just stand there and take it! You've got to hit her back! She won this round!" But knowing, from past experience, that failing to conserve energy will tire a fighter in seconds, to me this was a strategic move. A tired fighter is a beaten fighter. I won this fight by knockout in the second round. When my opponent had exhausted her supply of energy, when her strikes had lost their sting, when her moves went into slow motion and her hands dropped to her waist, I had only to unleash a couple of good right hands. What enabled me to do this was the fact that I had conserved energy while she had not.

How do you conserve energy, then? How do you keep from getting tired?

What is power?

Energy is what enables you to do work. If you have a lot of energy, you can do a lot of work. *Work* is defined as *force times distance*, **W=fd**. Work is the product of the component of force that acts in the direction of motion and the distance moved. Two things are of importance every time work is done:

- You must exert a force.

- Something must be moved by that force.

As an exercise in power, I had one of my students throw rear leg round house kicks in rapid succession on a kicking shield. I then had him throw lead leg round house kicks in rapid succession. Even though he had been throwing rear kicks first, and he was already getting tired, he told me how much easier it was to throw the lead round house fast.

For a given force, you can use a *shorter* distance to do *less* work. Work takes energy, so if you can do as little work as possible and still attain a considerable amount of power, you will last longer during heavy physical exertion. Because work is the product of the force and the distance, for a given amount of work, if you decrease the distance, you must increase the force, and vice versa.

If you view Mike Tyson's hooking strategies, you will find that he has tremendous power. One of the reasons is that he uses short, explosive moves that employ a very short distance for the exertion of the force. By relying on the power in his body, he can sacrifice distance for a greater advantage. Again, this becomes a gain/lose situation. We have talked about how distance allows you to build momentum for a strong punch; that's why your rear techniques are often stronger than your lead techniques. Increasing the distance, however, also increases the **WORK**load. If you continuously throw your strikes from a long distance, more work is done, and more energy is needed. This will tire you faster. Short punching, relying on the mass and explosiveness in your body, will help you conserve energy for a longer duration.

If you have a lot of energy, you can do a lot of work. There is *potential energy*, which is stored and held in readiness. Once released, this energy becomes *kinetic energy*, which we will look at in more detail in the Chapter 8. *Power* is a measure of how fast the work is done. **Power = work/time**. Being *twice as powerful as your opponent* means that you can do the same amount of work in half the time, or twice the work in the same time. Consider how much more tired you get when doing something fast than when doing it slow. Why is it easier to walk a mile than to run one? How about sprinting?

So, why do you get more tired when running a mile than when walking it? Why do you get more tired when fighting aggressively, than when relying on defense and only exploding with an offensive technique when it is to your advantage to do so?

The physics of the one-inch punch

One of the things that Bruce Lee was famous for was the one-inch punch. From what we have learned about power so far, the one-inch punch contradicts the principle of distance: *the longer the distance, the more time you have to build momentum for power*, which is also why boxers and kick-boxers fight with their stronger side farther from their opponent.

Many years ago, I heard about a martial artist of the Wing Chun system who could throw the one-inch punch. I looked him up and asked if he would demonstrate it to me. To my delight, he said he would. He handed me a phone book to hold against my chest for protection. He faced me, held his hand in the vertical position, and touched his finger tips to the phone book. He concentrated for a few seconds, then pulled his fingers into a fist. The next thing I knew, I was with my back against the wall five feet behind me. I was still holding the phone book, but my body was tingling from the force of his punch. The Wing Chun stylist told me that the explosive power comes from the elbow.

Explosiveness means higher *impulse*; a very short time at the moment of impact, and therefore a lot of power. Explosiveness can offset the lack of distance. From what I observed of the one-inch punch, the Wing Chun stylist's elbow was bent prior to throwing the punch. It was also held in front of his body along his centerline. In addition, his hand was in the vertical position, allowing him to keep his elbow down and use his body, which is heavier than the arm, to initiate the movement of the punch. I have experimented moderately with the one-inch punch and found that, even from a horse stance, you can attain considerable power by keeping your elbow in front of your body and pivoting your foot, hip, and shoulder in unison. But the initial move must be explosive. If the punch results in a push, or anything that even slightly resembles it, at that short a distance, not much power can be produced.

To every action there is an equal and opposite reaction. When one object collides with another, the total momentum is conserved, but may be *redistributed* to the other object. Consider a fist (with the full weight of the body behind it) moving toward a person at rest. When the fist hits the person, the momentum is redistributed to the person, and the person is moved back. **Momentum = mass X velocity**. If I had been proficient at the one-inch punch and the situation had been reversed, the Wing Chun stylist would also have been moved back by the force of my punch, but perhaps only half the distance that I was moved back, because he weighed so much more.

So, does the one-inch punch exist? Yes, it does. Can anybody do it? Not without a considerable amount of practice. Can a small person throw a powerful one-inch punch? Yes, but if his initial move is no faster than the heavier person's, then the heavier person's punch will have more power than the lightweight's.

Simply *knowing* what the principles of physics are, will not make you a great martial artist. It still takes years of training to get your body mechanics right, so that you can utilize these principles to their fullest. I think the easiest way to learn the one-inch punch is to work on synchronizing your moves, so that all parts of the body that are involved in throwing the punch can act as a unit.

Conservation of energy

There are many ways to conserve energy throughout a fight. One way is by relying on defense while your opponent is relying on offense. Another way is by using your *opponent's* movement to your advantage. The less you have to move, the more energy you will conserve. This is especially important when in a confined area, such as a boxing ring. Even in a street encounter, you will usually be in a confined area: a room in a house, a hallway, a parking lot, a car. Learning to control that confined area, and making your opponent move around you, will help you conserve energy longer than your opponent. Remember the saying: *with sufficient stamina, you can wear down the most powerful adversary.* Even if your opponent has the upper hand initially, if you can tire him out during the course of the fight, you can often turn the fight to your advantage. You should therefore strive to dominate the center, and make your opponent expend energy by moving around you.

When you get really tired, something as simple as taking a step forward may seem difficult. But since your forward motion adds momentum to your techniques and therefore power, you don't want to sacrifice it entirely. *Timing* is now crucial. If you can time your strikes to your opponent's forward motion, you will gain considerable momentum by making your opponent walk into your strikes. You will simultaneously conserve energy by minimizing your own movement.

Blocks are effective defensive tools against strikes or kicks. However, when defending against an attack, you are momentarily tying up the blocking weapon. This makes you unable to use it for offense until defense is complete. If you can rely on movement instead of blocking (slipping, bobbing and weaving, side-stepping, etc.) you will leave your hands and feet free for offense. In other words, you can use defense and offense simultaneously. While your opponent is still thinking of offense, you are already landing your next strike, while at the same time having successfully defended against his attack.

If your opponent likes to sweep, you can draw a sweep by giving him your leg and use the momentum and energy of your opponent's sweep to launch a counter-attack.

Try drawing a sweep by giving your opponent your leg. Energy is conserved by going with the motion of the attack, creating offense by bringing maximum power with least effort.

224   Fighting Science

When your leg has give, it enables you to go with the motion of the sweep. The power is redistributed over a longer time, and the impact of the attack will not be as damaging (**concept of impulse**). Because your opponent is on one leg and in the process of sweeping, he is also unable to move away until the sweep is complete. By utilizing the momentum of your opponent's sweep to set your own kick in motion, you will have accomplished three things: defended against the attack, conserved energy, and come back with offense when your opponent is the most vulnerable.

Conservation of energy during a fight that is lengthy may be crucial to winning. In the stand-up arts, energy can be conserved by dominating the center of the ring. This means very small movements on your part, with your opponent moving around you in big circles. If your opponent is cornered and you only give him one way out, he *has* to take that route. This makes fighting predictable, but to your advantage. If you know beforehand which way your opponent is going to move, you can launch an attack in that direction and land it with certainty.

In the pictures below, Martina has her back against the wall. If she moves to her right, her opponent could successfully land a round house kick or a knee strike. Martina will be walking into the power of the kick, the kick will land with certainty, and her opponent will conserve energy by utilizing as little movement as possible.

Cindy conserves energy by placing her opponent against the wall and giving her only one way to escape; to the right (the corner is to the left).

Let's look at the spinning back kick, and see how conservation of energy can make you land this relatively lengthy technique with power.

Using the spinning back kick strategically

Because a fight is dynamic, your opponent will seldom stand in one place and allow you to bombard him with punches and kicks. In order to land a technique with power, you must adjust to your opponent's movement. This is perhaps especially important when throwing a technique that is lengthy (takes more time than other techniques), or one that requires you to turn your back toward your opponent. The reason that the spinning back kick is difficult to land is because we fail to adjust to our opponent's movement.

If both you and your opponent are in left fighting stances, the best time to throw the spinning back kick is when your opponent moves to his *right* (your left), as this enables you to kick with less than 180 degrees of turn. In other words, your opponent will be walking into the kick (economy of motion). It is more economical to throw a kick which your opponent meets halfway, than to chase your opponent with the kick. Landing the spinning back kick when your opponent moves to his *left* (your right) is more difficult, because you will need to spin more than 180 degrees.

$$\text{Work} = \text{force} \times \text{distance}$$
$$\text{or}$$
$$W = Fd$$

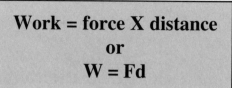

Work = **F**d

When your opponent moves to your left, you can land the spinning back kick with less than 180 degrees of turn. Your opponent will be walking into the kick.

Work = F**d**

When your opponent moves to your right, you must spin more than 180 degrees. You will be chasing your opponent with the kick.

Caution: Your opponent is more likely to move to his left than to his right, because it is easier to step left from a left fighting stance than to step right from a left fighting stance. You must therefore train for your opponent's lateral movement to his *left* (your right). The opposite is true from right fighting stances.

Conservation of Energy

The concept of adjusting to your opponent's lateral movement can be likened to shooting a bird in flight (from a perpendicular angle). Since the bird is moving forward, and you know that it will take some time for the bullet to reach the bird after you pull the trigger, you must aim slightly in front of the bird.

When throwing the spinning back kick against an opponent who is moving, you must aim your kick where he *is going to be* when the kick lands, and not where he is at the initiation of the kick.

The spinning back kick can be thrown at many different stages during the fight:

- As a follow-up off of another strike or kick.

- As defense against an aggressive fighter coming toward you.

- As a strategic move to keep your opponent on the ropes.

Because we know that it is easier and more economical to throw the spinning back kick when your opponent moves toward your back (toward your left if you are in a left fighting stance), by being strategically smart, you can now lure him into moving to your left. A good time to do this is when he is on the ropes. Being on the ropes is usually seen as a position of weakness, and your opponent will look for an escape path. By cutting off his escape path to your right, his only way out will be to your left. Now, when you can anticipate his move, landing the spinning back kick is almost a given. Because your opponent will be moving toward the kick and into its path of power, the kick will land regardless of how far your opponent has stepped when you throw the kick.

Summary and review

A body that is moving has a great deal of inertia, which makes it difficult to stop. If you can add your opponent's momentum to your own, you will assist the speed of your technique. This is why the soft arts are often referred to as yielding; you utilize your opponent's force against him.

Conservation of energy when striking

Energy can be conserved by doing as little work as possible. Work in physics is defined as the force times the distance.

$$\text{Work} = \text{force} \times \text{distance}$$

In essence, this means that if you utilize a lot of movement (a long distance), then, for a given force, you will be doing a lot of work, and therefore expend a lot of energy. This is why we must learn to utilize our opponent's work to our advantage. Work can also be thought of as power times time.

$$\text{Work} = \text{power} \times \text{time}$$

For a given amount of work, the shorter the time, the greater the power. In simple terms, this means that if you can do a lot of work in a short time, you have a lot of power.

Inertia is important to conservation of energy, because every time you have to overcome inertia, you expend energy and get tired. You should try to use both your own and your opponent's inertia to your advantage.

Conservation of energy:

1. You can conserve energy by using as little movement as possible, and relying on your opponent's movement. An example would be to dominate the center of the ring, forcing your opponent to move around you.

2. It is more economical to strike when your opponent is moving into the strike's path of power than when he is moving away. When throwing a spinning strike, think about the direction of the strike, and whether it will be more economical to strike when your opponent moves to your left or right.

3. If you miss with the spinning back fist, try to take advantage of your momentum by launching another strike with a looping or spinning path. Stopping the momentum and starting it again takes energy because you must overcome inertia.

Your opponent's inertia:

1. Going with the motion of a strike or block enables you to set your counter-attack in motion. You are now feeding off your opponent's energy. For example, if your opponent attempts a sweep, utilize the spinning motion to launch a spinning back fist.

2. If our opponent pushes you, try going with the motion of the push, or side-step and rely on his forward inertia to make him walk into your strike.

When your opponent pushes you, go with the motion and rely on your opponent's forward inertia to bring power to your strike or kick.

230 Fighting Science

Your own inertia:

1. Allow strikes to ricochet off the target. This helps you change direction easier, and you will expend less energy. Rather than using your muscular effort to bring your strike to a stop, use your opponent's body.

2. Leaning into a strike will starve the strike of power. Leaning also makes it difficult to snap the strike back after impact. Because your body is going forward, you must work against the weight and inertia of your own body. If you don't snap the strike back, it will result in a push and power will be lost.

Leaning into a strike will starve the strike of power. This is because the strike will be working against the inertia of your body.

Conservation of energy when kicking

Kicking takes more energy than striking. This is because the legs are heavier than the arms. Energy, when kicking, is conserved by relying on principles of physics that allow you to use as little effort as possible. Energy is also conserved by kicking when your opponent is moving into the kick's path of power. This allows you to use less of your own momentum to achieve an overall stronger momentum.

Body mechanics for energy conservation:

1. The side kick takes less effort to land when your opponent is slightly toward your back rather than your front. This is because the kick is usually thrown with your lead leg, with your opponent lined up with your hips.

Conservation of Energy 231

When your opponent moves toward your back, a spinning back kick is economical, because he will be moving into the kick's path of power.

2. The spinning back kick should be thrown when your opponent is moving into its path of power. Energy is conserved by spinning less than 180 degrees. If you spin more than 180 degrees, you will be chasing your opponent with the kick.

3. If your opponent moves toward your front, it would be economical to throw a round house kick.

If your opponent moves toward your front, a round house kick is economical, because he will be moving toward the kick's path of power.

4. Energy can be conserved by dropping an axe kick on your opponent's guard and bringing his upper body forward and into your follow-up strike.

Inertia in kicks:

1. Energy can be conserved in all kicks by chambering the leg prior to kicking, and by keeping the kick lined up with your centerline. There is less inertia in the leg the closer the weight is to the pivotal point.

2. A spinning kick that is not compact is difficult to accelerate, and has more rotational inertia than a kick that is tight.

3. Low kicks use less energy than high kicks, because they don't have to counteract gravity. Kicking to the legs also requires less movement, and is therefore economical.

Conservation of energy in defense

Defense, by itself, is a reactive mode that uses energy without gaining a tactical advantage. When your opponent strikes, he has a lot of inertia and is difficult to stop. Energy in defense can be conserved by going with the motion of your opponent's strike or kick, allowing his momentum to accelerate your counter-attack. Energy in defense can also be conserved by using the minimum amount of movement in your blocks, and allowing your opponent's strike to come to you rather than reaching for it. If your opponent's strike is forceful, you may want to redirect it rather than block it.

Using your opponent's inertia:

1. Use defensive motion to set up a counter-attack (inward forearm block to set-up a spinning back kick, for example). Because the motion of the inward forearm block is lateral and pivots your body, the spinning back kick is a logical follow-up. Any move in the same direction conserves energy.

2. Aside from moving the head or body, you can also use a full body pivot to avoid a strike or kick. This is done by rotating your body around its vertical axis. Energy is conserved because you avoid expending it in the block. You can also incorporate a step simultaneous to pivoting off the attack line.

Conservation of Energy 233

A pivot or step off the attack line conserves energy, because you are going with the motion of your opponent's strike and avoid expending energy on blocking.

3. Body rotation can also be used in conjunction with an offensive move, like a grab. The idea is to allow your opponent's energy to set your own body in motion.

When your opponent grabs you and pulls, go with the motion and launch a counter-attack.

4. When you are in a clinch and your opponent tries to wrestle you, go with the motion rather than resisting. How does this allow you to manipulate yourself into a better position? Can you use your opponent's momentum to your advantage?

5. Techniques that are not met with power (a parry or other redirection of the strike, or a defensive move like slipping) allow you to use your opponent's momentum to your advantage. When his strike misses, he still has inertia and will place himself in an inferior position.

6. Apply the principles of inertia and conservation of energy to defense and offense simultaneously (pick and counter, block and strike, bob and weave with a counter, slip and use opponent's inertia and momentum against him).

Conservation of energy in throws and takedowns

A properly executed throw will seem almost effortless. The effectiveness of a throw can be increased through the use of circular momentum. If the throw is linear in nature, it will result in your opponent going over the top; you will be carrying his weight and will expend a lot of energy.

Conserving energy through positioning:

1. Your first priority should be positioning. A superior position will restrict your opponent's fighting ability and better your own. The best position is generally behind your opponent's back.

2. Rather than moving yourself to the superior position, energy can be conserved by moving your opponent to the inferior position. For example, use his inherently weak neck to turn him with his back toward you.

Controlling your opponent's neck preparatory to a takedown enables you to move him into an inferior position without expending a lot of energy.

Conserving energy through movement:

1. A joint lock is often used preparatory to a takedown. Energy in a joint lock is conserved by moving your opponent close to your center of mass, and relying on short and quick moves to initiate the takedown.

2. Energy in a throw is conserved through the use of short and quick moves, using your opponent's momentum against him. There should be no stop in momentum, or you will expend energy on overcoming inertia.

Conservation of energy when grappling

Because of the proximity of the fighters in a grappling situation, you can conserve energy by using much of your opponent's movement or body weight against him. Although many people think of grappling as wrestling, strikes and kicks can also be used on the ground, along with larger moves, as when repositioning from controlling the feet to controlling the head.

The use of fine motor skills and your opponent's momentum:

1. Conserve energy through your opponent's struggle. If you are the heavier fighter, try to maintain the top position, making your opponent struggle under your weight. If you are lighter than him and end up on the bottom, rather than fighting against his weight, try to use joint controlling techniques combined with leverage.

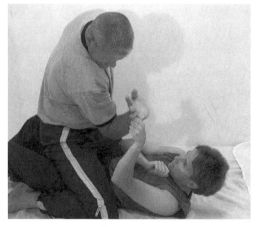

When you are on the bottom, a finger lock takes very little energy, yet it has shock value and is extremely effective.

2. It is important to keep a close proximity to your opponent, especially if you are in the inferior position. Once your opponent reacts to the joint lock, allow his heavier mass to initiate the move. Then roll with him to a more favorable position.

Staying close to your opponent allows you to rely on your opponent's energy expenditure to reverse positions.

Combining moves:

1. The fewer moves you make, the less energy you expend. Try to stay one step ahead of your opponent, and attempt to combine two moves into one. For example, taking your opponent to his stomach from a four-point stance, and moving his arm into the desired position simultaneously will conserve energy because you are combining the moves.

An elbow strike and a joint lock are two techniques that can be used simultaneously to conserve energy.

2. When your opponent is straddling you, controlling his head through torque and pain compliance will conserve energy.

Energy quiz

1. How can you keep from getting tired when fighting?

In a lengthy fight, it is almost impossible to keep from getting tired, even if you are in very good shape. However, a fighter can last longer during heavy physical exertion, if he learns how to conserve energy throughout the fight.

2. Some of the principles of physics and power seem to contradict and rely on a gain/lose situation, where sometimes you will need to give something up in order to gain a greater advantage. Explain how distance can work either to your advantage or disadvantage.

The longer the distance, the more time you have to build momentum. This is why your rear techniques often seem stronger than your lead techniques. But increasing the distance also increases the *workload*. When more work is done, more energy is needed, and you will get tired faster.

3. What is the main advantage of power?

The main advantage of power is the acceleration it can produce. Acceleration, in turn, translates into more devastating strikes.

4. Name a few ways in which you can conserve energy.

Energy can be conserved by shortening the movement required to execute a technique. For example, use short explosive moves that rely on the power in your body. Or get your opponent to expend more energy than you by dominating the center of the fight and making him move around you. When you get tired, rely on defense until you have recuperated enough to come back with good offense. A fighter who is very tired, yet tries to rely on offense, will lose the sting in his strikes.

5. For power, when is best to throw the spinning back kick and why?

The best time to throw the spinning back kick is when your opponent moves toward its path of power. Because the spinning back kick relies on circular movement, and because a fight often moves circular, you should strive to throw the kick whenever your opponent moves in a direction that enables you to spin less than 180 degrees. This will help you conserve energy and time through less movement.

6. How can you use the spinning back kick as a strategic move, forcing your opponent to move into its path of power?

If your opponent is in a left fighting stance, you know that he will favor moving to your right (his left). This is because his left foot is forward, making it easier to step with that foot first (the opposite is true if he is in a right fighting stance). Because you can conserve more energy in the spinning back kick if your opponent moves to your left, you can use the kick strategically against an opponent on the ropes by cutting off his escape path to your right, thereby forcing him to move to your left.

Glossary

Energy--That which enables you to do work. If you have a lot of energy, you can do a lot of work.

Force--Any influence that can cause an object to be accelerated.

Impulse--Change in momentum, or the force times the time interval. The shorter the time, the greater the force, and vice versa.

Kinetic Energy--Half the mass times the velocity squared. Kinetic energy depends on the mass and speed of the object. If a fighter can double his speed, he can quadruple his kinetic energy. Kinetic energy has a great capability of doing damage.

Mass--The quantity of matter in an object. When acted upon by gravity, we can use mass interchangeably with weight.

Momentum--The product of the mass of an object and its velocity. The heavier the object, and the faster it travels, the greater the momentum.

Potential Energy--Energy that is stored and held in readiness. Once released, it becomes kinetic energy.

Power--In physics, power is equal to work divided by time interval. Cutting the time interval enables you to attain more power. The martial artist often thinks of power in terms of how much damage one is able to do when landing a strike.

Velocity---A measure of the speed of an object and its direction.

Work--Force times distance. It takes energy to do work. You can conserve energy by using a shorter distance to do less work.

Ki-netic Energy
Mind Over Matter

Fact: You *will* get hit some time during your martial arts career. Knowing this in advance, which fighter would you rather take a punch from? The big fat one, or the small skinny one?

From what we have learned about body mass in motion, it would seem logical to choose the small fighter over the big one. But because power is a combination of many factors, including momentum, distance, speed, acceleration, inertia, and energy, sometimes one of these factors must be decreased in order to increase another. Consider inertia. ***Inertia*** means resistance to change. This also means that once an object is set in motion, it is difficult to stop it or to change its direction. But it is also difficult to set the object in motion in the first place. The more ***mass***, the more inertia. A heavy fighter must therefore expend more ***energy*** than a light fighter to set himself in motion.

Because ***force*** is a combination of mass and acceleration, $F=ma$, we can assume that it is more damaging to get hit by a heavy fighter who has the mass of his body behind the strikes. But this is true only if the fighter also has the ability to ***accelerate*** as fast as a lightweight. It is interesting to note that it is more damaging to get hit by a fast moving lightweight than by a slow moving heavyweight. Why? Because a lightweight moving with the same momentum as the heavyweight has more **kinetic energy**.

You cannot strike somebody effectively without setting your fist (or striking weapon) in motion. Energy of motion is called kinetic energy. Kinetic energy depends on the mass *and* speed of the object. We already know that the more mass, the more capable you are of producing a powerful strike. We also know that the greater the speed, the more capable you are of producing a powerful strike.

Note: A fast moving heavyweight can produce more power than an *equally* fast moving lightweight, because the heavyweight has more momentum. But fast moving heavyweights are not often seen. Because of inertia, it takes more energy to set a heavyweight into motion than a lightweight. When you expend more energy, you get tired quicker. And when you get tired, your strikes will lose their sting.

The power paradox

The interesting point about kinetic energy is that it is equal to *half* the mass multiplied by the *square* of the velocity. **Kinetic energy = ½ mass times the velocity squared**. A light fighter who is only half as massive as a heavy fighter, but who is moving with the same momentum, will have twice the velocity (speed). But in the kinetic energy equation the velocity is squared. This means that if the speed of an object is doubled, its kinetic energy is quadrupled, and the object can do *four times* as much work at only twice the speed. An object moving twice as fast as another takes four times as much work to stop. Faster speed = more kinetic energy *by the square* (not proportional). While momentum is proportional to velocity, kinetic energy is proportional to the square of velocity. An object that moves with twice the velocity of another object of the same mass has twice the momentum, but four times the kinetic energy. It can provide twice the impulse, but do four times as much work. The object will penetrate four times as far. It will deliver four times the damage!

Speed now becomes extremely important. Twice the velocity gives you four times the kinetic energy. If a fighter is only half as massive as his opponent, however, twice the velocity won't give him four times the kinetic energy, because his mass is half, but it will give him *twice* the kinetic energy. It takes more effort to move a heavy fighter, so which is best: a heavy fighter with a lot of momentum, or a lighter fighter with a lot of kinetic energy? This is where speed will offset weight, and being a lightweight can actually give you the power advantage.

Consider a side kick. This kick is often used to keep an aggressive opponent at a distance.

When the kick connects, the *0* is transferred from your body to your opponent's body. If you can speed up the kick, so that it strikes with twice the speed, you will be able to knock your opponent twice as far. Or if you can time the kick so that your opponent steps forward and into it, you will add his momentum to that of the kick, and the kick will knock him farther than if you rely solely on your own momentum. So, by increasing the momentum, you will increase the ***impulse***.

A side kick thrown twice as fast has twice the impulse, but *four times* the kinetic energy. It will penetrate four times as far and do four times the damage! Increasing the speed increases the impulse *proportionally*, but the kinetic energy is increased *by the square*. The heavyweight, who is twice as heavy as you, can generate twice the momentum and twice the impulse. But if you can increase your speed to twice that of the heavyweight, you will not be able to do equal damage to him as one might think, but twice the damage! If you can rely on "bouncing" (reversal of motion), you will increase the impulse even more. If the kick can bounce (be thrown back) elastically with no loss in speed, the change in momentum and impulse is doubled. To give your strikes penetrating force, the energy must be focused over a small surface area.

So, the momentum is increased by increasing the speed, and by keeping the weight of your body behind all strikes. But if the product of your opponent's mass and velocity matches yours, you will be stopped short. The jarring exerted from the combined momentum of you and your opponent will be the same. Being able to stop a fighter from pushing you back against the ropes is one thing. But again, if you are the lightweight, and you can move faster than your opponent, you will do more damage, because a light fighter moving with the same momentum has more kinetic energy. Instead of trying to gain weight for an upcoming fight, try to gain speed.

How important is kinetic energy?

Kinetic energy is energy in motion with penetrating capabilities. If I were to tell you that a punch has more momentum than a bullet, would you believe me? The heavier something is, the greater the momentum. One of the best examples of this is a cruise ship. Because of its enormous size, the captain may have to turn the engines off a mile or more from shore, or he would be unable to bring the ship to a stop. Your fist might weigh 20 times more than a bullet, so the momentum of your fist is great compared to the bullet, even at a slower speed. But the kinetic energy of the bullet will offset the advantages of momentum. Let's figure this out:

Momentum = mass (weight) X velocity (speed)
Kinetic Energy = 1/2 mass X velocity squared

Let's say that the mass of the fist is 2 kg (for simplicity). The speed of the fist is 3 meters/second (for simplicity). Using the momentum equation above, the momentum of the fist is **2 X 3 = 6**.

Let's say that the mass of the bullet is 0.01 kg. The speed of the bullet is 200 meters/second. Using the momentum equation above, the momentum of the bullet is **0.01 X 200 = 2**.

So, the fist has more momentum than the bullet. You can knock somebody over with your fist, but the bullet is more likely to penetrate the person's body. Let's see how much kinetic energy the bullet has.

KE (fist) = 2/2 X 3 squared = **1 X 9 = 9**.
KE (bullet) = 0.01/2 X 200 squared = **0.005 X 40000 = 200**.

The ratio of kinetic energy between the bullet and the fist is **200/9 = 22.22**.

The bullet has **22.22** times the kinetic energy of the heavier fist, and is 22.22 times more likely to do penetrating damage! These numbers do not represent the actual weights and speeds of the fist and bullet; I just used numbers that seemed

Ki-netic Energy 245

reasonable and were easy to work with. However, the result is true to life, and is proven through the equation. The point is that unless your goal is to knock somebody over, or to use your weight in grappling, weight does not necessarily do a lot of damage. Speed does.

However, kinetic energy does not travel through air easily, because the air molecules would use the path of least resistance and disperse to the sides. It is therefore *not* possible to strike a person from across the room without touching him. However, it is possible to send kinetic energy through a stack of bricks and break the one on the bottom while leaving the others intact, but *only* if the bricks are touching and the energy has a medium to travel through.

Many board and brick breaking demonstrations use spacers to separate the boards. When this is done, you cannot choose to break the bottom board only, while leaving those on top intact. Unless the top block breaks and touches the one below it, the energy does not have a medium to travel through.

What does the space between the boards or bricks do? My husband, Tom Sprague, can tear a two-thousand page phone book in half with his bare hands (and not at the crease in the middle either, mind you!) relying only on the principles of physics. At first sight, this may seem like an impossible task. However, if you can create a tiny bit of space between each page, the pages will no longer act as a unit, and it will take only a little more strength tearing the whole phone book in half than tearing just one single sheet of paper.

Try this: Grab the closed phone book with both hands and place your thumbs in the middle. Make a kink where your thumbs are by bending the book slightly toward you. Then bend the book back into its original shape while simultaneously exerting pressure against the sides of the book. This will create air between each page, but not enough to be visible to the onlookers. Providing that you can tear through the cover, the rest of the task should be easy and can be done in one quick move, giving the appearance of tremendous strength.

This exercise does require that your hands are big enough to reach around two-thousand pages and get a good grip. Practice on a thinner book first.

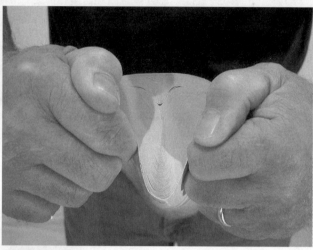
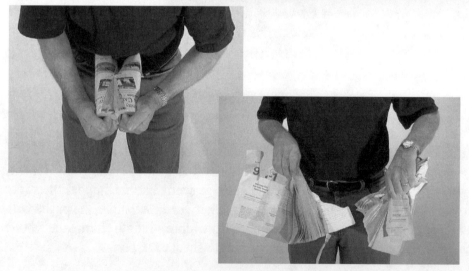

Tom Sprague tearing phone book in half.

When there is space between each brick, you only need to exert a force that is a little larger than what it takes to break one brick. When this brick collapses on the one below it, momentum and energy is transferred down through the pile. A small amount of the force is also lost due to friction and heat.

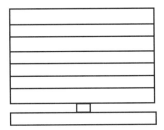

Here is a stack of eight bricks. The top seven bricks are resting on one another. The one on the bottom is separated from the others by a layer of air. However, there is a small peg leading from the top stack to the brick on the bottom. If the karate practitioner can strike the top of the pile with enough force to break *one* brick only, the energy will transfer through the pile, then through the peg, and down to the bottom brick, which will break. The top seven bricks will not break, because they act as a *unit*, and there is not enough power in the practitioner's strike to break the whole unit. If there was no peg (only air) between the unit and the bottom brick, the energy would not have a medium to travel through, and it would not be possible to break the brick on the bottom.

A bullet can penetrate a target easier than a fist, not only because of the speed and kinetic energy, but also because the force is focused over a very small surface area. A spectacular martial arts demonstration involves bending steel rods pushed against the practitioner's throat. Undoubtedly, being successful at this takes some tightening of the muscles in the throat, and choosing an area of the throat that is less sensitive to damage (probably not the Adam's apple).

The rods are placed at a slight diagonal angle upward against the throat and not straight into it. This allows the practitioner to spread the force over a larger surface area. Once the rods have started to bend, a weakness is created in the rods, and focus can be maintained downward to finish the exercise. The principle that is used is ***pounds per square inch*** (force per surface area). Next time you see this demonstration, ask yourself why the rods are so long? If the practitioner were to use a foot long rod instead, would he be as successful? The longer the rod, the more sag you get in the middle, which benefits the performer. **Warning: Do not try this exercise at home or without proper supervision. Severe injury may result!**

Mind over matter

The breaking of boards and bricks is often demonstrated in karate tournaments. One of these demonstrations involves not the actual power of a person's fist or foot as it goes through the target, but the "internal power" of the practitioner. One demonstration involves a bed of nails and a stack of bricks that is broken with a sledge hammer against the practitioner's body. This is really a demonstration of the difference between momentum and kinetic energy. Remember, momentum creates wallop (the ability to knock something over); kinetic energy does damage (penetrates).

When the hammer is swung, it has both momentum and kinetic energy. The bricks break, but the martial artist who is resting on the bed of nails is not harmed. Yet, every bit of the momentum of the hammer goes into the artist's body. However, most of the kinetic energy never gets to him, but is absorbed into the breaking of the bricks. The energy that does remain is distributed over hundreds of nails. The force per surface area is not enough to penetrate the skin. If this same experiment were done lying on *one* nail only, I guarantee that the results would be different! **Don't try it!**

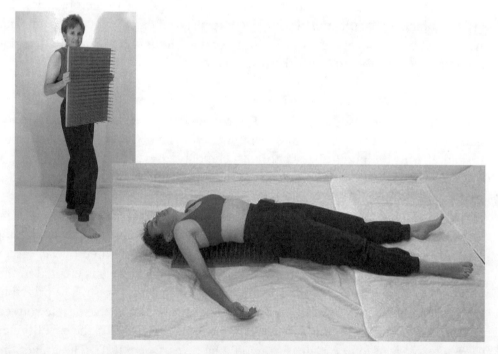

Martina is resting on a bed of 352 nails. The great number of nails makes it sound especially bad. But the more nails, the less force per square inch, and the less likely that a nail will penetrate the skin.

Next time you witness this demonstration, pay attention to *how* the practitioner is lowered and raised from the bed of nails. Most likely, he will not sit and lie on it the way one would normally sit and lie on a bed. Instead, he will have help from one or more fellow martial artists. This enables him to touch all the nails simultaneously, already from the beginning distributing the force over the whole area of his back, with no single nail given the "advantage" of the practitioner's whole weight.

And then there are those martial artists who walk barefoot on burning coal as part of a belt promotion. Yes, it takes great fortitude to overcome the fear of getting burned. But if the martial artist knew something about physics, he would have little to fear. Walking barefoot on burning coal is possible not because of the martial artist's extensive training in overcoming pain, but because your feet are protected by natural perspiration, which creates an insulating layer between the coals and the skin of the feet. Nervousness before performing the feat is actually to the practitioner's advantage, because nervousness causes more perspiration.

Have you ever extinguished a candle by wetting your fingers and squeezing the wick? Same principle.

So, what about this "mind over matter" type of fighting mentioned at the beginning of the book, where a martial artist claims to be able to strike somebody from across the room without being within reach to physically touch him? Does it really work?

I once asked to get this demonstrated, but was told that the energy was too great, and that it would have a very damaging effect on me. And because "true" martial artists don't use their powers to hurt, but only in self-defense, the master was not willing to demonstrate his technique on me. So, I asked if he would consider striking an object instead of a person, say, for example, to knock a vase off a table? I was unsuccessful in this matter as well. I was told that the object has to be *alive* for it to work. Why? After a lot of thought on this matter, I came to the conclusion that it is in fact possible to strike a person from a distance without physically touching him, but only if that person knows in advance that he is going to get hit, hence why the object has to be alive; a dead object does not have the ability to know in advance. This theory works not on the actual powers of the master, but on the *perceived* powers of the master, and on the *anticipation* of the person receiving the blow, and on the principle of freezing.

If you have ever been in a full-contact match, where you know beforehand that you will get hit, and where you also know that you will get judged on your ability to take a good punch, and in particular if your opponent has a reputation, this anticipation of the blow is likely to cause you to freeze (tense) whenever you think that you are going to get hit, even if your opponent does not throw an actual strike. For example, if you have already tasted some of his power and you know that it hurts, and therefore you respect it, or if you respect this person because of his reputation, he now simply needs to fake a strike to get you to tense. This tensing often translates into pain, although it has no lasting effect.

Flash knockout

I have seen boxing matches where one of the fighters has been accused of going down with a flash knockout (where the strike that hits the fighter appears nearly powerless). Some think that such matches are fixed, where the loser has been instructed to act as if he is getting knocked out. My belief is that those fighters who do get knocked out by a flash knockout do so because of their fear and anticipation of the other fighter's power, and not because they were actually hit by a powerful strike. This is where the real mind over matter comes in, and where the master's skill lies in intimidation rather than in *ki* or other physical principle. This is also why a fighter must know in advance that he is going to get hit. Lacking this knowledge would make him unable to fear the strike, and he would therefore not get tense in anticipation of it.

It is, of course, possible to get knocked out by a powerful strike. But once you know that you are going to get hit, you must train yourself to eliminate the fear factor or the desire to tense. When I start training with a new student, one of the exercises we do in the second or third lesson is getting used to seeing a punch coming at you. I put my boxing gloves on and have my student move with me, but without throwing anything. I will now randomly start flicking jabs toward the student's face, but without actually hitting him. Our natural reaction to this is to close our eyes and jerk our head back. After sufficient training, the student will learn to stay focused on the opponent, and not close his eyes. Once you get used to the fight game and your confidence grows, you can use your own "powers" (your mental determination) to counteract your tendency to freeze. This is when the fighter, or master, who has relied on his reputation will be stripped, and this mind over matter fighting will no longer have an effect.

Ki-netic Energy 251

Other incredible physical feats

As mentioned earlier, kinetic energy must have a medium to travel through. That's why it is not possible to break the bottom brick in the pile, unless that brick is connected somehow to the top bricks. That is also why it is not possible to strike somebody from across the room without being within reach to physically touch him. But what about air? Does air not qualify as a medium for the energy to travel through?

Air *does* qualify, but the problem is that air is a gas that is *compressible*. If possible, the air molecules will also disperse to the sides, taking the path of least resistance. For air to be used as a medium for kinetic energy, it must be compressed at a very fast rate. If a person stands at one end of the room and mentally focuses on hitting a person at the other end of the room, nothing will happen. If he throws a punch in the air toward the person at the other end of the room, again, nothing will happen. However, if you throw a punch that comes fast enough and close enough to the target, it is possible to move the hairs on your opponent's head without actually touching. If the air molecules can be compressed fast enough to exert pressure in the direction of the punch, it will be felt by your opponent. However, it will not hurt him, because the human body is not capable of generating a rate of compression that is fast enough to keep the air molecules from dispersing to the sides. (A jet-engine, on the other hand, can generate a very strong and focused stream of air. I have personally seen a pickup truck rolled on its side, when the driver drove behind a running jet engine at a distance of over a hundred yards. All the windows in the truck were also shattered).

Some martial artists are able to extinguish a candle with a punch, without touching the flame. When it comes to air and fire, an interesting thing occurs. Fire needs air in order to burn, because air fuels the fire. Too much air, however, will extinguish the flame. If the practitioner threw his fist more slowly and provided a gentler draft, the candle would burn more violently. By using speed and snap, the air molecules will hit the flame in greater number over shorter time, and the cooling effect is enough to extinguish the candle. If you were to extinguish a candle by blowing, which works best: blowing slowly and gently, or blowing forcefully in a quick spurt? If the practitioner's fist didn't get close enough to the candle, he wouldn't be able to extinguish it, because the air molecules would disperse to the sides. This exercise takes both speed and distance awareness, but is relatively easy and can be achieved with very little practice.

252   Fighting Science

The more massive an object, the more inertia it has. This means that if the object is in a state of motion, it will be difficult to stop it. It also means that if the object is in a state of rest, it will be difficult to set it in motion. In the picture below, Martina is cutting a board with a knife hand shuto. Her strike has a lot of kinetic energy.

$$\text{Kinetic energy} = 1/2 \text{ mass} \times \text{velocity squared}$$

$$\text{Kinetic Energy} = 1/2 m V^2$$

Martina cuts a board with a shuto (knife hand strike).

Some demonstrations involve cutting the top off a free standing bottle. Here, several things occur simultaneously:

- The bottle has a wider base than top. This means that the base is stable with a low ***center of gravity***.

- Because of the shape of the bottle and the liquid in it, the bottom is heavier than the top and therefore possesses a great deal of inertia. If the bottle was empty, it would need to be tied to the table or be held steady by someone.

- The speed with which the break occurs is important. More speed means more kinetic energy, and it is therefore possible (in this case, literally) to take the strike through the target.

- Because of the speed, there is more kinetic energy than momentum. Kinetic energy does damage, while momentum provides wallop. If the momentum was greater and the kinetic energy less, the top would not be cut off, but the whole bottle would be knocked over instead.

- Using the side of your hand enables you to focus the energy over a small surface area, which in turn means more force per square inch. If you use the whole palm of your hand instead, you would not be as likely to succeed.

The strength of the arch

Some martial artists give the appearance of making themselves lighter by walking on fragile objects without breaking them. But because we are acted upon by the same amount of gravity all the time, the only way you can make yourself lighter is by going on a diet and losing weight. However, a fragile object (a glass, for example) may become strong because of its structural shape. Why is the roof over an athletic arena dome shaped and not flat? You obviously can't have pillars in an athletic arena to support the weight of the roof, so a dome is built. The strength of a dome follows the same principles as the strength of an arch (the St. Louis Gateway Arch, for example, or a doorway arch which has to support the weight of the building without collapsing).

A flat roof will sag in the middle. Take a sheet of paper and hold it horizontally at each end. Observe how the middle is sagging. Now apply pressure inward (toward the center) until the paper becomes arched upward. These inward forces compress the building materials, increasing the stability. You can even place an object of considerable weight on top of the arched paper without it sagging.

Another point to consider is how the force is distributed. Most people would agree that an egg is fragile. But if you take an egg and make your hands conform to its shape, you will find that the egg is remarkably strong. Breaking the egg by squeezing your hands together, applying pressure equally across the whole surface area of the egg, will make the egg difficult to break. If you put one finger on each end of the egg, however, and poke it, the force is focused over a very small surface area and will penetrate the egg easier. When you see a martial artist walk on a fragile object, it is likely that the object is dome shaped, and that the martial artist's foot is made to conform to the shape of the object, distributing the weight over a larger surface area.

There is one more thing, which is a bit outside of the physics of power, but which still needs to be said, and that is the *capability of the human mind*.

Imposing your will

When you cut through the hype, fighting is a simple game, where the fighter who imposes his will wins. In other words, make your opponent fight *your* fight. This should be your foremost thought and strategy both during your fight preparation and during the actual contest. A stance, for example, should be both offensive and defensive. Offensively, you must give your opponent a threat: look confident and ready. Defensively, you must cover your openings. A stance can therefore be thought of as both mental and physical. Don't let your mind use its psychological powers to drain energy from you and slow your speed.

In the picture at left, note how Martina's legs are slightly bent to lower her center of gravity for stability. Yet, she is not rigid; she is not just standing there waiting to see what will happen. Through her stance, she has already taken charge of the fight. Note also how her rear foot is ready to push off against the floor for a quick forward shuffle. This readiness adds to the mental characteristics of the stance. It is obvious that she can't wait to engage in battle.

> Next time you fight, you're going to win because you are ready. Because the power is there, because the skill is there, because the wind is there, and because the attitude is there. And I'm going to stand in your corner and watch you eat your opponent alive and spit out the bones.

Summary and review

While both momentum and kinetic energy rely on mass and velocity, in the kinetic energy equation, velocity is stronger than mass. This means that an object that is lightweight but fast will have more kinetic energy than one that is heavy and slow. Because kinetic energy depends so much on speed, it applies mostly to the striking arts, and perhaps especially to those fantastic breaking demonstrations you see in tournaments. In this summary and review section, I will refrain from categorizing kinetic energy into sub-sections on striking, kicking, defense, throws, and grappling, like I have done in previous chapters. Instead, let's review the most important points about each physics concept we have learned. Keep in mind that most of these concepts tend to overlap, that there are tradeoffs and payoffs, and that it is difficult to work with one concept exclusively.

Center of gravity:

1. Without balance, your strikes, kicks, defenses, throws, takedowns, and grappling techniques will be nearly worthless. Balance may therefore be the most important of all concepts.

2. Balance relates to your center of gravity, and is generally increased by keeping a low and wide base. But there are some drawbacks. A base that is too low or too wide decreases your mobility.

Momentum:

1. Momentum is a combination of the mass of an object and its velocity. Thus, a heavyweight tends to produce more momentum than a lightweight. Because the human body is only capable of moving so fast, mass seems to be more important than speed. Momentum can be achieved by placing your body mass behind all strikes.

2. Momentum is generally used to move an object through a distance, as when knocking your opponent back or down.

Direction:

1. Velocity involves not just the speed of an object, but also its direction. If your body mechanics contradict, there will be a conflict with the resultant force, and maximum power cannot be attained.

2. Moves can be broken down into vectors (arrows symbolizing the direction and strength of the move). If two vectors point in different directions, the force will split. This is true whether you use a strike, a takedown or throw, or a grappling technique.

Rotational speed and friction:

1. The higher the speed, the greater the power. Great power can be built in spinning techniques, because the technique is allowed to travel through a greater distance on the rotational plane. The farther the strike is from the center of rotation, the greater the speed.

2. Friction acts in a direction opposite of motion, and is less in air than on the ground. That's why techniques involving a jump can be accelerated faster. Another reason is because a jump allows you to contract your body to decrease the inertia and rely on conservation of angular momentum.

Impulse:

1. Impulse involves how quickly the momentum comes to a stop. In the striking arts, force is increased by relying on speed and snap, with the momentum reversed through the opposing surface. In defense, force is decreased by going with the motion of your opponent's strike.

2. In the grappling arts, force is increased by slamming your opponent straight down into the ground, or by striking a body part that is in contact with the ground. Forcee is decreased by sequentially absorbing the shock of a fall.

Conservation of energy:

1. Energy is conserved by using as little movement as possible. Because momentum comes from movement of mass, short explosive moves may be more energy conserving than longer moves.
2. Energy is also conserved by relying on your opponent's motion, allowing him to walk into your strikes or kicks.

Kinetic energy:

1. The faster the speed of your strike or kick, the greater the kinetic energy. A lightweight fighter, who has less inertia to overcome, can generally throw faster strikes and kicks than his heavier opponent.

2. The smaller the impact weapon, the greater the force per square inch. For example, impacting with the ball of your foot, rather than with the whole bottom portion, will increase the force.

Kinetic energy quiz

1. Explain how speed can outweigh the benefits of mass.

It takes a lot of energy to set a massive object into motion. When you expend energy, you get tired. A lightweight fighter can therefore be quicker than a heavyweight, without sacrificing as much energy. Twice the speed increases the kinetic energy by the square.

2. Why is kinetic energy capable of doing so much damage?

A strike that is thrown twice as fast as your opponent's strike, will have twice the impulse, and twice the ability to move your opponent back. But it will have four times the kinetic energy, and will penetrate four times as far, and do four times the damage. In the kinetic energy equation, the speed is squared. Mass only provides wallop (the ability to knock something over); kinetic energy does damage.

3. Why is it beneficial to use a striking weapon that employs a surface area that is as small as possible?

Penetrating force is produced best when the force is focused over an area as small as possible, because the force per square inch will be greater. This can be demonstrated using a board and your bed at home. First, place a large wooden board on top of your mattress. Then, climb up on the bed and stand on top of the board. Even though the mattress is soft, you will notice hardly any indentation at all from the weight of your body. Now, remove the board and again climb up on the bed. Stand on one foot only. Is there greater penetration when your weight is spread over the area of your foot only, than when it is spread over the whole area of the wooden board?

4. What are the advantages/disadvantages of gaining weight before an upcoming fight?

If you're fighting in weight classes, it might be to your advantage to be at the top of your weight class. A few extra pounds will give you the ability to move your opponent back, especially when he is tired. If there are no weight classes, and the range in weight between fighters is great, it may be to your advantage not to gain too much weight. The extra weight makes you slower, giving a lighter fighter the advantage *kinetic energy-wise.*

5. What effect does "imposing your will" have on an opponent?

The mind is a very powerful tool. Having confidence and belief in yourself will often give you more respect than perhaps you are worthy of. An opponent who has superb technical skill and experience may lose a fight before it has even begun, simply because your confidence strips him of his own.

6. Is it possible to get knocked out by your opponent's strike, even if you are imposing your will on him?

Of course! If your opponent's strike is powerful enough, just having confidence and "thinking" that you will not get knocked out is not enough. But confidence sure helps!

Glossary

Acceleration--Changes in speed and/or direction. An object that is in motion *and* changes its direction (a car driving up the clover leaf on-ramp to the highway), will accelerate even if there is no change in speed. You can feel that acceleration is taking place by the way your body lurches forward, back, or sideways.

Center Of Gravity--The point on an object where all its weight seems to be focused. An object of uniform shape and weight has the center of gravity in the middle. An object of non-uniform shape and weight has the center of gravity toward the heavier end. To remain stable, the center of gravity should be as low as possible, and above the foundation.

Energy--That which enables you to do work. If you have a lot of energy, you can do a lot of work.

F=ma--The force is equal to the mass times the acceleration. The more massive a fighter, and the more he can accelerate, the more force he can produce.

Force--Any influence that can cause an object to be accelerated.

Impulse--Change in momentum, or the force times the time interval. The shorter the time, the greater the force, and vice versa.

Inertia---Resistance to change in motion. An object at rest tends to stay at rest; an object in motion tends to stay in motion.

Kinetic Energy--Half the mass times the velocity squared. Kinetic energy depends on the mass and the speed of the object. If a fighter can double his speed, he can quadruple his kinetic energy. Kinetic energy has a great capability of doing damage.

Mass--The quantity of matter in an object. When acted upon by gravity, we can use mass interchangeably with weight.

Mind Over Matter--This type of fighting exists only because we allow our minds to get tricked by the anticipation of a strike.

Momentum--The product of the mass of an object and its velocity. The heavier the object, and the faster it travels, the greater the momentum.

Pounds Per Square Inch---The narrower the base of an object, the more pounds per square inch. A narrow base is less stable than a wide base, but it also has more penetrating capability. When bending steel rods pushed against the throat, the practitioner will be more successful if the force is spread over a large surface area.

Conclusion

Throughout my years of study in the martial arts, what has intrigued me the most is perhaps the many contradictions. In modern day language, martial arts means *the intricate study of combat*. Because combat means battle, struggle, conflict, it seems logical to think of a martial artist as a person who has studied conflict and is skilled at combat. The more traditional way of interpreting the term martial arts is through *budo*, which means "the *way* of the warrior." Aikido founder and master Morihei Ueshiba emphasized that "a martial art must be a procreative force, producing love, which in turn will lead to a creative, rich life." (The Spirit Of Aikido--by Kisshomaru Ueshiba, Kodansha International) Most martial arts instructors drill into us from the beginning that we are only to use the art when we absolutely have to. And the philosophies of the arts often cater to love rather than war.

When looking at power and the principles of physics, I have found great many paradoxes as well; the gain/lose situations I have been mentioning throughout the book, where sometimes you must give something up in order to gain a greater advantage. For example:

- When pinning your opponent to the ground, it is more difficult for him to throw you off if you spread your weight as much as possible. On the other hand, using only a very small part of your body when pinning (the elbow or knee, for example) will make you less stable, yet it enables you to increase the intensity of the technique by placing more pounds per square inch on the target.

- It is better to be heavy than to be light, because more mass means more force behind your strikes. But more mass also means more inertia, and more difficult to set in motion. Without motion, there is no power.

When working with the martial arts, you must tilt the *law of averages* (the probability) to your advantage. When rolling a die, for example, what is the probability that you will roll a six? Because there are six sides to the die, the probability is one in six, which means that if you were to roll the die six times, you are likely to roll a six one of those times. If you roll the die twenty-four times, you are likely to roll a six four of those times. In pure probability, as in the case with the die, not much can be done to affect the law of averages. But what if you were to take a multiple choice test, with four possible answers to each question? If such a test were taken blind folded, the probability that you would score correctly would be one in four. Without any knowledge whatsoever of the questions being asked, you would still come through the test with a score of 25%. If you needed a score of 75% to pass this test, you had only to increase your knowledge by 50% and not by 75%, as one might have thought before taking the test.

Before any action at all is taken, you should look at the probability of a successful outcome. Because of their athletic build, speed, and flexibility, some people appear to be naturals. The person who is not a natural can still come out on top by evaluating his situation beforehand, and using opposing qualities to exploit his opponent's strengths. In The Ultimate Fighting Championship I, Sumo wrestler Teila Tuli (410 lbs.) met Savate fighter Gerard Gordeau (216 lbs.). About twenty seconds into the fight, Teila charged forward. While backpedaling, Gerard threw several punches to Teila's head that did little or no damage. It was obvious that the almost twice as heavy Teila had the momentum and power advantage. However, shortly thereafter, Teila lost his balance and went down. Although Teila had several seconds to get back to his feet before Gerard could close distance, *because of the inertia on 410 pounds of weight*, he was unable to get up in time. Gerard threw a powerful round house kick to Teila's head, which knocked a tooth out and cut the Sumo wrestler under the eye. This is a fine example of how the advantage of weight can be exploited by a lighter opponent. The heavier a fighter, the more momentum he can produce, *but* the more energy he will have to expend getting back to his feet.

What makes physics so appealing is that it is not really up for debate. The principles of physics have been established through the *scientific method*, where scientists recognize problems, form a hypothesis (an educated guess), and perform experiments to test the hypothesis. Once the outcome is established, a theory is formed. When a hypothesis has been tested over and over without being contradicted, it may become known as a law or *principle*.

The principles of physics apply to all people at all times, regardless of which art you study. Understanding physics therefore allows you to understand power, speed, and endurance as a whole, and to work these principles to your benefit. Size and weight, for example, may seem like an advantage, but it is also a limitation. Whether the glass is half-full or half-empty depends on what's in the glass, and on what you are trying to achieve. It is my hope that this book will help you use physics to turn your special limitations, whatever they may be, into strengths.

Good Luck!
Never Quit!
Never Give Up!
Never Say Die!

Author Bio

Martina Sprague started training in the martial arts in 1987. She holds black belts in kickboxing and modern freestyle, and has an extensive background in Ed Parker's system of Kenpo karate. She grew up in Stockholm/Sweden, and came to the United States as a student in high school in 1980. She later attended Westminster College in Salt Lake City, and earned a Bachelor of Science Degree in Aviation. She has been a flight instructor on light aircraft since 1986, and currently works for Delta Air Lines, where she is responsible for weight and balance calculations and proper loading of aircraft.

Martina's extensive experience as an instructor, both in aviation and martial arts, has helped her acquire in-depth knowledge of the principles of teaching and learning. Her main focus is on helping students refine their martial arts skills through the use of the laws of physics.

Index

A

acceleration 78, 127-128, 143-146, 148, 150, 153, 172, 189, 241, 259
accuracy 95
adding momentums 83, 99, 100, 124, 157, 189
aggressive opponent 95
aikido 193
angled attacks 171
application 21
axe kick 107

B

back fist 104
back kick 112, 158
balance 39-40, 47-50-59, 64, 79, 87, 102, 157, 173, 255
basic movement theory 40, 79
blocking 66-68, 91, 112-115, 166-167, 170, 223
board breaking 198
bobbing and weaving 68-69, 129, 146, 176
body throw 73
budo 261

C

center of gravity 41-51, 58-59, 79, 87, 93, 150, 252-259
center of mass 41
chambering 65, 179, 232
circular motion 50, 72, 76, 116, 120, 128, 130, 139, 142, 148, 171, 176, 184
circular takedown 182
combinations 22, 25, 30, 63, 96, 110, 130-132, 140,-141, 144-147, 149, 176, 179, 200
completion of motion principle 27
conservation of angular momentum 169

conservation of energy 222-228, 256
correlation 21
counter-attack 28, 112, 114, 211, 137, 232
crescent kick 106, 111
crouching 92

D

defense 32-3, 66, 88-91, 112, 146, 181, 211, 232
direction 101, 121, 127-133, 150, 157, 160
distance 90,-94, 118, 128, 138-139, 149, 167-170, 180, 220
distraction 121
double end bag 130
downward kicks 107
downward strike 105

E

economy of motion 96, 130
elbow strike 101, 104, 114
energy 220, 239, 241, 259
escalation 25-26
evasive movement 68
explosiveness 34

F

$F=ma$ 153
falling 194
figure four 121
flash knockout 250
force 15, 41, 127, 153, 166, 192-193, 202-204, 217-220, 239-241, 259
force per square inch 42, 79,109, 120, 124
forward roll 194
four-point stance 236
friction 155-159, 189, 256
front kick 106, 109, 135
full armbar 121

G

grappling 74, 118-121, 149, 184, 213-214, 235
gravity 52, 79, 84, 124, 139
guard 97

H

half-stepping 94
hammer fist 105
hand speed 96, 97
hard style 49-50, 79
heavy bag 132
hip throw 74, 119
hook 130, 137, 138
horse stance 39, 40, 46, 62, 79

I

impact 192-196, 201-204
impulse 15, 97, 124, 192,-198, 200-207, 217, 222-224, 239, 243, 256, 260
indomitable spirit 36
inertia 52, 79, 82, 124-128, 148-150, 153, 160-161, 179, 189, 200, 217, 228, 260
internal energy 191

J

jab 101, 130, 134, 161-162
joint lock 73-74, 147-149, 197, 212-213, 235
judo 193
jump kick 107, 110, 156-159
jump spinning back kick 158-159

K

kenpo 199, 200
ki 191, 217
kicking 63-65, 106, 110,-111, 134, 143, 155, 179, 201, 209,-210, 230-232
kinetic energy 15, 124, 128, 153, 192, 217, 221, 239, 242-244, 252-260

knee strike 109-111
kung-fu 49

L

lateral movement 227
law of averages 261
law of non-resistance 35
lead leg kick 47, 107-108
leverage 24, 30, 164, 183-184
linear movement 71
linear speed 168, 189
linear strikes 49
looping kicks 106
looping strikes 104-105

M

mass 32, 79, 82-86, 101, 124, 127-128, 153, 160, 166, 189, 217, 239-241, 260
mechanics 21
mind over matter 260
mobility 71
momentum 15, 29, 32, 47, 56, 79-89, 97,-101, 124, 128, 137, 153, 157, 167, 178, 189, 192, 204,-207, 217, 221-222, 239, 255, 260

N

neck throw 73
Newton's First Law Of Motion 160, 189
Newton's principles of motion 155
Newton's Second Law Of Motion 166, 189
Newton's Third Law Of Motion 166, 189

O

one-inch punch 163, 221-222
optimal reach 90
orward roll 193
overhand strike 138-140

P

pain compliance 26-28

palm strike 104
parrying 50, 68, 115
penetrating force 166
pivoting 68, 93-94, 113
positioning 28, 234
potential energy 221, 239
pounds per square inch 194, 217, 247, 260
power 15-19, 84,-89, 95-98, 128,-134, 221, 230, 239-242, 256
projectile 139, 153
punching 93-101, 130-134, 163, 193, 205
pushing impact 203, 217
pushing off 93

R

range 90, 149
reach 90-94, 108, 168
relaxation 131-132, 192
resultant 85, 124, 130, 133, 139-146, 153, 162
rhythm 96
ridge hand 104-105
rotational inertia 85, 124, 171-175, 217, 201
rotational speed 56, 80, 168-169
round house kick 106-108, 132-133, 231

S

sensory overload 50, 80, 97, 124
sequencing of techniques 36
shattering impact 202, 217
shin block 67
shoot 120
shotokan 49
shuto 105
side break fall 195
side kick
   106, 109, 112, 135, 158, 160, 243
slipping 68-69, 129
soft styles 49, 50, 80
speed 86-89, 101, 127-131, 160-167, 173-176, 242-243
spinning back fist 172-173, 177, 200, 208-229
spinning back kick 108, 134, 174, 225-227, 231
spinning strikes 172, 177-178, 200

splitting the resultant 140-148, 150, 162
stability 40
stamina 219
stance 60, 72
stick 169-170
stinging impact 202, 217
stomp 107
straddling 121, 150
straight kicks 106
strategy 23, 35
strength 19
strike
striking 62, 101-103, 140, 166-167, 171,-176, 197, 207, 214, 229-230
sweeping 66, 74, 165, 223, 224

T

tackling 118
takedown 54-57, 74, 117-120, 136, 147-148, 165, 182-185, 197, 212, 235
throwing 50-51, 73, 116-118, 147-148, 182, 193, 212, 235
timing 94, 223
torque 15, 24, 80, 163-165, 181-186

U

understanding 21
uppercut 129, 130, 138
upward strikes 105

V

vectors 57, 80, 85, 124, 130-134, 142, 153, 162, 256
velocity 86, 101, 124, 127, 139, 153, 239, 255

W

weapon 169-171
weight 19, 82, 89, 95, 101-102, 131
work 15, 239

Also Available from Turtle Press:

Guide to Martial Arts Injury Care and Prevention
Solo Training
Fighter's Fact Book
Conceptual Self-defense
Martial Arts After 40
Warrior Speed
The Martial Arts Training Diary
The Martial Arts Training Diary for Kids
Teaching Martial Arts
Combat Strategy
The Art of Harmony
A Guide to Rape Awareness and Prevention
Total MindBody Training
1,001 Ways to Motivate Yourself and Others
Ultimate Fitness through Martial Arts
Weight Training for Martial Artists
A Part of the Ribbon: A Time Travel Adventure
Herding the Ox
Neng Da: Super Punches
Taekwondo Kyorugi: Olympic Style Sparring
Martial Arts for Women
Parents' Guide to Martial Arts
Strike Like Lightning: Meditations on Nature
Everyday Warriors

For more information:
Turtle Press
PO Box 290206
Wethersfield CT 06129-206
1-800-77-TURTL
e-mail: sales@turtlepress.com

http://www.turtlepress.com